SCIENCE FOR SALE

SCIENCE FOR SALE

How the US Government Uses
Powerful Corporations and Leading
Universities to Support Government
Policies, Silence Top Scientists,
Jeopardize Our Health, and Protect
Corporate Profits

David L. Lewis, PhD

Foreword by
Marc A. Edwards, PhD

Afterword by
Caroline Snyder, PhD

Skyhorse Publishing

Skyhorse Publishing books may be purchased in bulk at special discounts for sales promotion, corporate gifts, fund-raising, or educational purposes. Special editions can also be created to specifications. For details, contact the Special Sales Department, Skyhorse Publishing, 307 West 36th Street, 11th Floor, New York, NY 10018 or info@skyhorsepublishing.com.

Skyhorse® and Skyhorse Publishing® are registered trademarks of Skyhorse Publishing, Inc.®, a Delaware corporation.

www.skyhorsepublishing.com

10 9 8 7 6 5 4 3 2 1

Library of Congress Cataloging-in-Publication Data is available on file.

ISBN: 978-1-5107-4310-6
Ebook ISBN: 978-1-5107-4317-5

Printed in the United States of America

Author's Proceeds go to the Nonprofit
National Whistleblower's Center (www.whistleblowers.org)

This book is dedicated to my wife, Kathy, the love of my life,
and to our children, Josh and Jedd, our greatest joy.

CONTENTS

CONTENTS

FOREWORD

"Mayor: The matter in question is not a purely scientific one; it is a complex affair; it has both a technical and an economic side....As a subordinate official, you have no right to express any conviction at odds with that of your superiors.

Dr. Stockman: What I am doing, I am doing in the name of truth and for the sake of my conscience"[1-2]

—Ibsen, An Enemy of the People, 1882

S cientific progress and the advancement of civilization are inextricably linked. At the threshold of the third millennium the virtuous cycle of public support and investment in science, which has yielded multifold returns via improved quality and quantity of life, has been eroded at a most inopportune moment—sound science is the best hope for saving humankind from itself.[3] Yet the public is rightly becoming skeptical of the motives of modern science, which has frequently betrayed the public trust and faces rising external criticism to reform.[4-6]

Ibsen's classic "An Enemy of the People," illustrates the dangers of consensus science under the influence of institutions. Heroic Dr. Stockman, who sacrifices his career and personal standing to protect the innocent from harm, discovers that the scientific truth offends self-serving bureaucrats and powerful economic interests, and is soon denounced as a lunatic and "An enemy of the people" for defending the truth. Unfortunately, scientists have been very poor students of their own history, and suffer from misplaced moral overconfidence in their enterprise, institutions and judgment, which have always been subject to fraud, misconduct and the delusion of crowds as chronicled brilliantly by Ibsen.[7] More recently, the inexorable rise of big and collaborative science, necessary to solving increasingly complex problems, has created a "belongingness" imperative for scientists on a scale and intensity never imagined by Whyte in "The Organization Man."[8] Success in modern science has become too dependent on networking and teamwork, appeasing group pressures in the name of cohesion and expedience—scientific dissent and skepticism that play critical roles in good science are rarely encouraged or even tolerated.

The experiences and career of Dr. David Lewis, like the fictional Dr. Stockman, exemplify the type of heroic action that will be necessary if modern science is to become "self-correcting." His work provides important insights to those who care about the scientific enterprise, and is a call to action for brave souls capable of sacrificing to preserve the integrity and promise of science for future generations. David highlights a pernicious threat to science which he first defined and personally named, "institutional scientific misconduct," in which science has little to do with the seeking of truth and serving the public good, but rather, is conscripted to perpetuate the power/policies of institutions via conjuring of pseudo-scientific illusions. He defines institutional scientific misconduct as follows:[9-10]

> *Institutional Scientific Misconduct or, in the case of research, Institutional Research Misconduct, is the fraudulent manipulation of science by government agencies, corporations and academic institutions to support government policies and industry practices. It often involves the suppression of credible scientific research by using false allegations of scientific or ethical misconduct against honest scientists who document the adverse effects of government policies and industry practices on public health and the environment.*

Institutional scientific misconduct has gone largely unappreciated and unreported because its practitioners are often the very agencies we have empowered to police scientific integrity in one way or another. And, at least superficially, they appear to lack a direct profit motive that would explain or cause scrutiny of their unethical actions. Many still mistakenly believe that such motives are a necessary inducement to incentivize scientific misconduct. In my opinion, the abuses and dangers of institutional scientific misconduct far exceed those arising from misconduct in industrial science, as society has developed certain checks and balances to control industrial science abuses while still preserving its undeniable benefits to society.

I personally witnessed institutional scientific misconduct during an event known in the press as "The Washington D.C. Lead in Drinking Water Crisis," during which government agencies first inadvertently triggered release of hazardous levels of lead from plumbing to drinking water in tens of thousands of homes in the nation's capital, and later criminally hid the health threat from the public for more than three years.[11-18] Unaware of the health hazard in their drinking water, parents were unable to protect their unborn fetuses, children or even themselves from the best known neurotoxin (lead). Hired to develop an understanding of the problem by the United States Environmental Protection Agency, and then acting to hold the public's interest paramount over my own self-interest or that of the agencies, my funding and access to data from experiments I designed were terminated.[13]

After the dimensions of the hazard were exposed in an award winning *Washington Post* investigative report in early 2004, the agencies and their minions published a series of reports relying on false or non-existent data, which created an illusion that the unprecedented exposures and the agencies' criminal actions had not poisoned a single man, woman or child.[19-23] The fact that their results contradicted over two thousand years of human knowledge and experiences regarding health harm from elevated lead in drinking water, was neither deterrent or hurdle, in their exercise of power via abuse of science.[23] The agencies then stood by and even encouraged the spread of their pseudo-scientific lies, harming children in other parts of the United States and the world. None of those responsible were held accountable, although a few were later inexplicably given "Gold Medals" and other honors for public service by the agencies.[23]

It took more than six years of personal effort and a U.S. Congressional investigation to finally discover that the agencies' "data," which they purportedly relied on for their landmark conclusions and publications, had strangely disappeared if it ever existed in the first place.[20-23] Other dimensions of their falsification were also exposed.[20] Yet, to this day, agency reports and publications for which data cannot be found, and with conclusions proven to be false by numerous investigations and some of the agencies' more recent research, have still not been retracted.[20-23] The Centers for Disease Control even issued a new falsified report , which in an Orwellian fashion, brazenly claimed that their prior work had actually concluded that D.C. citizens had been poisoned by the years of high lead in water, but that the CDC had communicated their results poorly and had been misunderstood by readers![24-26] Demands to retract the falsification of a falsification, which e-mails demonstrate was crafted at the highest levels of the CDC, were completely ignored.[27]

I mention my own experiences because they are relatively well-documented, the information has been thoroughly vetted in the press and in Congress, and before living through it myself I would never have believed it possible.[26, 28-29] Having grown up worshipping at the altar of science, I never even thought to question the motives of government agencies, the policemen that we pay to protect us, and who would seem to have no financial motive to behave unethically. Indeed, when I first heard of David Lewis and his battles with the EPA in the 1990s, I was too busy establishing my own career to pay any attention, much less delve into the details and sort out the truth. That is the reality of scientific whistleblowing. Whistleblowers and heroic scientists, are always destined for singular journeys, because even our most supportive colleagues have little choice but to be bystanders and there is no higher authority in science that can serve as an impartial judge and jury. As David found, sadly, a courtroom run by non-scientists is often the only recourse. It takes great courage, persistence and a supportive family to see matters through to a conclusion, as David did in his years as a "successful" EPA whistleblower.

As is his nature, above and beyond his documented successes in demonstrating shortcomings in CDC infection-control policies and exposing wrong-doing at EPA, David does not shy away from new, challenging and controversial subjects. The book details his personal journey to try and better understand details of the Brian Deer and Wakefield drama, a subject on which, after decades of painful struggle, establishment science has reached a consensus verdict that Wakefield was guilty of scientific misconduct. David's work and belief suggests that this consensus is an over-reach supported by dubious data and interpretations. As a scientist with a newly discovered appreciation of history, I can only say that there are many prior precedents that should give one pause whenever institutions harshly judge their critics, or those who are reporting results inconsistent with the political winds of the day. Our institutions, which fancy themselves as impartial, beyond reproach and worthy of passing judgment on others, never seem to find that their own egregious actions ever rise to the level of scientific misconduct.

Consider the Cyril Burt affair. Lauded for his work on genetics and intelligence before dying in 1971, Burt's work came under intense scrutiny shortly thereafter.[7] Unable to defend himself because he was dead, and a key critic had successfully recommended that all of his notes/data be burned, a series of articles concluded that Burt's career and life's work had been a complete fraud.[7] But later work noted that Burt's key results had withstood the highest test of science, replication by others with modern methods, while obtaining virtually identical results. The bizarre rush to impeachment included an apparently false claim that Burt had even gone so far as to invent the names of his collaborators and co-authors![7,30-32]   It seems safe to say that the Burt affair will never be satisfactorily resolved, even though his key results and conclusions ultimately withstood the test of time.

Unsatisfactory resolution of scientific controversy in cases of scientific misconduct and whistleblowing, is currently the rule and not the exception. I recently defined whistleblowing, as it is practiced in government agencies, as "a human perversion of natural selection, whereby weak, deceitful, cowards survive, and strong, honest heroic truth-tellers are destroyed."[33] David Lewis is one of a few whistleblowers who was not destroyed, who continued to practice science productively, and who remains fiercely independent and productive. He willingly shares his vast experiences and wisdom with others, as was the case when he spoke to my graduate engineering ethics classes, and by offering assistance to other whistleblowers so that their journey is not quite as lonely as it might otherwise be.

Despite experiences that would break a normal human, Lewis remains optimistic about the future of science, and is rightly proud of his career before, during and after his personal ordeals with institutional misconduct within EPA. These characteristics might go far in explaining his personal success as a scientist and citizen. As you read his story which is part self-reflection and part memoir, Lewis feeds and encourages critical debate on many subjects, but first and foremost he embodies Ibsen's observation that "The strongest man in the world is he who stands most alone."[33] I cannot help but reflect that if brave whistleblowers such as David Lewis were to ever go extinct, given the magnitude of the science and engineering challenges that confront us; our civilization is not likely to be far behind.[34]

Marc Edwards
Charles Lunsford Professor of Civil Engineering
Virginia Tech

ACKNOWLEDGMENTS

This book is the culmination of innumerable acts of kindness, moral support, and technical guidance extended to me by family, friends, supervisors, coworkers, and colleagues at the U.S. Environmental Protection Agency, Centers for Disease Control & Prevention, Food & Drug Administration, Harvard University's Science Technology and Society Program, Boston University's Department of Environmental Health, and to the University of Georgia (UGA) Department of Marine Sciences, School of Ecology, and College of Engineering—especially Professor David Gattie. Since publishing my 1996 *Nature* commentary, "EPA Science: Casualty of Election Politics," the National Whistleblowers Center (www.whistleblowers.org) was my refuge. So were a number of government administrators, political appointees and elected officials, and others, including Rosemarie Russo, former director of the EPA research laboratory in Athens, Georgia; Robert Hodson, former UGA director of marine programs; Bernard Goldstein, former EPA assistant administrator for the Office of Research & Development under President Reagan; Jerry Melillo, former associate director for the environment in the White House Office of Science & Technology under President Clinton; and the late congressman Charlie Norwood.

Stephen and Michael Kohn and others at the law firm of Kohn, Kohn, and Colapinto in Washington, DC, and Ed Hallman and Richard Wingate of Hallman & Wingate in Marietta, Georgia, carried me through my legal battles with EPA and the treated sewage sludge (aka biosolids) industry from 1996 through 2013. I'm also grateful to my attorneys, James Carter in Madison, Georgia, and Finis Williams in Concord, New Hampshire, and Ed Hallman for their help regarding efforts by Brian Deer, the *British Medical Journal* (*BMJ*), and their supporters in the biosolids industry to undermine my current research on environmental triggers associated with autism and other neurological and immunological disorders and diseases. Christopher Shaw at the University of British Columbia is my primary collaborator. Technical guidance and/or financial support provided by Caroline Snyder (www.sludgefacts.org), Barry Segal, Claire and Al Dwoskin, and Abby Rockefeller (www.sludgenews.org), have been also critical to my success.

For enabling me to write this book, I am particularly indebted to Sheldon Krimsky, Tony Lyons, and Kristin Kulsavage. I'm also very grateful to Marc Edwards and Caroline Snyder for writing the foreword and afterword.

Finally, my investigations into the original documentation behind allegations of research fraud against Andrew Wakefield would have never succeeded without the full cooperation of Andrew Wakefield and his wife, Carmel.

PROLOGUE

This book reveals in graphic detail how government control over the scientific enterprise, which President Dwight D. Eisenhower predicted would eventually pose a grave threat to America's future, has finally come to pass. It also builds upon the prolific work of authors Daniel Greenberg, Sheldon Krimsky, David Michaels, and others.

The American silver plug penny minted in the 1790s proclaims liberty as the parent of science and industry. The founding fathers were convinced that freedom from corruption was vital both to a healthy economy and to scientific progress.[1] Few people today would disagree, and recent, highly publicized events surrounding the collapse of the housing market, corporate fraud, and the dire need for campaign finance reform have made the public well aware of the alarming influence that corrupt special interests have gained over the political process in the last several decades. While their insidious effects on the economy are well documented, most people have only had a glimpse of their impact on science.

During my thirty-plus years as a research microbiologist in the Environmental Protection Agency's Office of Research and Development (ORD) and at the University of Georgia (UGA), I experienced the far-reaching influence of corrupt special interests firsthand. As this book will describe, my dealings with civil servants, corporate managers, elected officials, and other scientists expose the ease—and disturbing regularity—with which a small group of individuals, motivated by profit or personal advancement, can completely hijack important areas of research science at even our most trusted institutions. The result is that today, many government-funded scientific endeavors have become little more than an arm of industry marketing efforts and government policymakers.

Many factors lead to the ease with which the scientific process is corrupted by outside influences. Some are largely unavoidable. The complexity of the science itself often makes it so that only a tiny fraction of outside observers have the background knowledge to notice when the pieces don't add up to a whole. On top of that, research is expensive, time-consuming, and inherently risky, making it hard for the small number of people who are able to understand any given area of research to unveil the corruption.

But the culture of our scientific institutions, and the priorities of many of their leaders, shares a lot of the blame. A 2008 survey of ORD's scientists by the Cambridge-based Union of Concerned Scientists reflects what many of my colleagues have become accustomed to: researchers are systematically subjected to top-down pressure to avoid conducting research or drawing conclusions that undermine government policies.[2] In a great many cases, those who do are fired, have their careers dead-ended, and are sometimes even prosecuted and imprisoned. These problems are mirrored in industry, which hires scientists to support its business. And they have spread to universities, which are heavily invested in obtaining grants that ultimately support government policies and industry practices.

If the trend continues, integrity in science may one day become about as rare as a silver plug penny. Unfortunately, organizations dealing with scientific misconduct are designed only to weed out those who commit fraud behind the backs of the institutions where they work. But the greatest threat of all is the purposeful corruption of the scientific enterprise by leaders within the institutions themselves. The science they create is often only an illusion, designed to deceive, and the scientists they destroy to protect that illusion are often our best.

Throughout my career as a research scientist, I've worked in areas where policymakers and industrial managers have a keen interest in controlling what gets published in the scientific literature. I have watched government officials, university administrators, and corporate executives manipulate science without restraint time after time to advance and protect their own interests, funding scientists to carry out research projects with predetermined outcomes, fudging data, and using false allegations of research misconduct to eliminate scientists who question their "science."

Since 1996, I've spent much of my time fighting governmental, industrial, and academic entities jointly engaged in efforts to stop my research and discredit my coauthors and me by any means necessary. This book describes the most important issues that my coauthors and I have investigated, along with important research topics that leaders at government agencies and in the corporate world have prevented me from ever undertaking. Along the way, I've discovered much about the methods that are sometimes used within government agencies, corporations, and academic institutions to manipulate science.

My coauthors and I, for example, were the first researchers to document adverse health effects associated with treated sewage sludges (biosolids) applied according to EPA's current regulation, the 503 sludge rule. This rule allows municipalities to collect industrial pollutants at wastewater treatment plants and spread them on farms, forests, school playgrounds, and other public and private lands without monitoring any pollutants other than nine metals and two nutrients, nitrogen and phosphorus.

EPA's attempts to stop our research and discredit the researchers with false allegations of research misconduct prompted two congressional hearings by the House Science Committee, a review by the National Academy of Sciences, and the passage by Congress of the No Fear Act of 2002. At first, Democrats in the House of Representatives refused to support efforts to clean up the scientific fraud and the silencing of concerned scientists behind EPA's biosolids program, choosing instead to cast Republicans as anti-environmental for attacking EPA regulations.[3] Then, as a Senate Briefing was scheduled, Republicans torpedoed that effort.[4]

Similarly, I spent almost two years obtaining and analyzing the U.K. General Medical Council's (GMC's) confidential documents behind allegations of research misconduct that Brian Deer and the *British Medical Journal* (BMJ) published against Dr. Andrew Wakefield. In the process, I discovered a document showing that the analysis of patient records that Deer published in 2010 perfectly matches an analysis requested by the GMC's lawyers in the GMC proceedings four years earlier. The analysis, which Deer published in the BMJ, was the result of a deliberate plan by individuals working for the GMC to conflate a blinded expert analysis of biopsy slides with routine pathology reports to make it appear that Wakefield had misinterpreted the records to link the MMR vaccine to autism. What the GMC's lawyers could probably never get away with in the courtroom—which was to condemn Andrew Wakefield for research fraud—Deer accomplished by publishing the GMC's convoluted analysis in the BMJ.

My hope is that this book will give our judicial system, the news media, and the general public a better idea of what goes on behind the scenes, where enormous resources are being invested to create the illusion of science needed to protect government policies and industry practices. Somehow, we must find a way to prevent this illusion from supplanting the real science that is desperately needed to protect public health and the environment. It is up to us to ensure that future generations do not pay the price for the institutional research misconduct that has become such a large part of science during our generation.

PRESIDENT DWIGHT D. EISENHOWER'S FAREWELL ADDRESS

President Dwight D. Eisenhower on Scientific Research:[1]

A steadily increasing share is conducted for, by, or at the direction of the Federal government.... The free university, historically the fountainhead of free ideas and scientific discovery, has experienced a revolution in the conduct of research. Partly because of the huge costs involved, a government contract becomes virtually a substitute for intellectual curiosity.... The prospect of domination of the nation's scholars by Federal employment, project allocations, and the power of money is ever present and is gravely to be regarded. Yet, in holding scientific research and discovery in respect, as we should, we must also be alert to the equal and opposite danger that public policy could itself become the captive of a scientific-technological elite.

STORY OF THE IRON HORSE

David Lewis. Photo courtesy of Walker Montgomery, circa 2010.

The famous Iron Horse is a symbol of the University of Georgia. It stands exiled to a cornfield visible from Georgia State Highway 15 south of Watkinsville, Georgia, near the author's home. Forged by artist Abbott Pattison and erected on UGA's main campus in May 1954, it's an example of iron sculpture introduced at the end of World War II.

Modern art was new to the campus, and students at the agricultural university abhorred it. As night fell, firefighters turned their hoses on a large crowd that piled hay and old tires around the Iron Horse and set it on fire. In a PBS documentary aired in 1980, Pattison said that he considered it a lynching. The author was similarly banished from the UGA campus in 2008.

SCIENCE
FOR SALE

ASK YOUR DENTIST

DO YOU HEAT-STERILIZE YOUR HANDPIECES AFTER EVERY PATIENT?

Everyone should have regular dental care. It not only saves your teeth; it may well save your life. As a microbiologist, I know how important it is to get regular dental checkups and maintain good dental hygiene. But, after conducting research on dental handpieces used to drill and polish teeth, I would never let a dentist or hygienist work on my teeth unless their handpieces have been *heat-sterilized after every patient*. Many don't, and you should always ask to make sure.

Key Players

Research conducted by the author and his coauthors at UGA, Washington University's Medical School, and Loma Linda University's School of Dentistry prompted the CDC, the FDA and other public health organizations worldwide to recommend heat-sterilization for every item that enters the oral cavity before it can be reused on another patient.[1] There were many key individuals and organizations in government, industry, and academia that fought on one side or the other. Here are some of the key players.

David Kessler, FDA Commissioner

Thomas Arrowsmith-Lowe, who handled dental issues for the FDA, walked into Kessler's office in 1992 with a copy of our *Lancet* study and the journal's editorial in hand. The editors summarized: "On p. 1252, Lewis and colleagues report that HIV-infected material can be sucked back into waterlines and expelled via a dental handpiece."[2] At a meeting at the headquarters of the American Dental Association in Chicago, Tom announced that the FDA was sending a letter to every dentist in the United States, and every possession of the United States, instructing them to heat-sterilize their handpieces after every patient.[3]

Harold Jaffee, CDC Director

In an interview with ABC *Primetime Live* producer Sylvia Chase in 1992, Jaffee watched a videotape of me operating a dental drill and "prophy angle" used for cleaning teeth after they had been exposed to blood and prepared for the next patient according to CDC guidelines. Traces of red blood could be seen streaming out as the devices were run over a container of clear water. "Is it not the same thing—this kind of blood transfer—as sharing a needle?" Chase asked.

Dr. Jaffee opened his mouth, but no words came out. After a long pause, he said, "Clearly, we don't want one patient to be exposed to another's blood."

Diane Sawyer introduced the segment by announcing that the CDC had decided to change its guidelines. Dr. Donald Marianos, the head of the CDC's dental section, called me the next morning to say what an impact the visual demonstration had on the staff at the CDC. Evelyn Lincoln, President Kennedy's personal secretary, also called me. She said that he would have personally taken action had this surfaced on his watch.

Kimberly Bergalis, University of Florida Student

Despite suffering in the final stages of AIDS, and unable to speak louder than a whisper, Kimberly testified before Congress. She waged a national campaign to force the government to get to the bottom of how she and at least five other patients in a Florida dental practice contracted HIV from their dentist—and stop it from ever happening again. Barbara Webb, a retired schoolteacher who was one of the other five, donated a blood sample for us to use in a study we published in *Nature Medicine*.

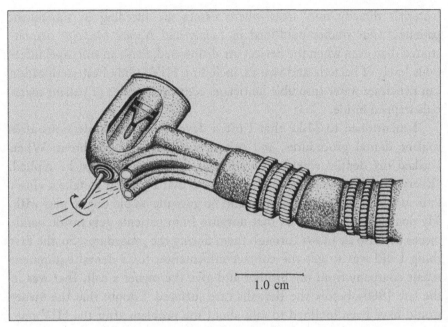

Dental handpiece drawing by David Lewis.

The Story

My older brother Mike joined the Navy in 1962, and was trained as a dental technician. After serving his four years, he worked for a dental supply company in Atlanta, Georgia. While visiting dental practices throughout the Southeast, he noticed traces of blood in the crevices of high-speed dental handpieces prepared for the next patient. Dental handpieces are divided into two categories: high-speed, for drilling, and low-speed, for polishing.

Dental drills run on air pressure controlled with foot-pedals. When the pressure is cut off, suck-back causes blood, saliva, and other patient materials to be drawn back into the handpiece. Although dental handpieces are re-lubricated between patients, they are not cleaned internally.

Slow-speed handpieces used to clean and polish teeth have the same problem. As hygienists scrape plaque from tooth surfaces along the gum line, bleeding occurs. Then, when low-speed handpieces equipped with prophy angles—rotating rubber cups—are run along the gum line, they suck back abrasive pastes contaminated with blood and saliva as the air-flow is disrupted. When reused, they expel the traces of blood and other patient materials directly into areas where tissues are bleeding in subsequent patients.[4] Our studies published in *Lancet* and *Nature Medicine* demonstrated that, even when the devices are disinfected, they can still expel infectious levels of bacteria and viruses, including HIV.[5] Only heat-sterilization can penetrate water-insoluble lubricants containing traces of patient materials trapped inside.

I mentioned to Mike that I often developed throat infections after routine dental procedures, and tested positive for strep throat. When I asked my dentist whether he autoclaved his dental tools, he replied, "Everything but handpieces." I told Mike that I wanted to take a closer look at what's inside handpieces, and he gave me an old one to play with. My dentist had assured me that nothing from patients gets inside handpieces because air blows through them during the procedures. So, the first thing I did was to get the contact information for a dental equipment repair company from my brother and give the owner a call. That was in the late 1980s, before the Bergalis case surfaced. I doubt that the owner would have been inclined to talk about the problem after the HIV outbreak caused widespread panic.

When I called the repair shop, I told them I was Mike's brother and wanted to do a little research on handpiece contamination. I asked, "What do you see inside them when they're repaired?"

He replied, "Tooth material, amalgam, bits of tissue, blood."

"Well, if it's going in," I said, "then it must be coming back out when the handpiece is reused, right?"

"That's right," he replied. To check it out myself, I scooped some of the debris from inside the old handpiece my brother gave me, and took a look at it using an electron microscope with X-ray diffraction to detect heavy metals. Lots of red blood cells, tissue, and mercury-silver amalgam particles were clearly visible.

It was unsettling to me, as a microbiologist, to realize what had been injected into my bloodstream with dental drills over the years. All I could think about was the countless strains of antibiotic-resistant staphylococci and streptococci from thousands upon thousands of patients lodging on my heart valves, and remaining there—a few cells here, a few cells there. Little ticking time bombs buried in some microscopic scar tissue or cholesterol deposit, waiting for my immune system to go downhill from diabetes or some other chronic disease, just sticking it out until I grow old. There's nothing in the world that can be done about it now.

I asked Mike if he knew any dentists who heat-sterilized their hand-pieces after every patient. He recommended Robert Boe, a dentist in the Atlanta area who was known for welcoming AIDS patients. Soon, I was taking my wife and two children to Dr. Boe to get our dental work done, and driving to his office after getting off work at EPA to experiment with dental handpieces. Dr. Boe was one of only 1 percent of dentists in the United States who was heat-sterilizing handpieces after each patient at the time.

HIV Outbreak in Florida

In July 1990, the Centers for Disease Control and Prevention (CDC) in Atlanta reported a possible case of HIV transmission in a Florida dental practice.[6] It would eventually trace the source of the infection in a University of Florida student, Kimberly Bergalis, and five other patients with no identified risk factors to their dentist, Dr. David Acer.[7] In 1991, Dr. Acer's hometown newspaper published a front-page story about our research, which ran in more than seven hundred papers nationwide.[8] One of our studies included a blood sample from one of Acer's patients.[9]

The CDC considered the possibility of Dr. Acer directly transmitting his HIV infection to six or more patients through needlestick injuries to be highly improbable. It also considered transmission by dental devices to be highly improbable, but wouldn't rule it out. The reasoning of CDC scientists was that HIV is highly sensitive to the 2 percent glutaraldehyde solution, which Acer's staff used to disinfect his handpieces and reusable prophy angles.[10] Moreover, a number of patients who contracted the dentist's HIV only had cleanings done; the CDC considered these procedures to be non-invasive and, therefore, unlikely to transmit HIV.

By exposing reusable prophy angles to blood from HIV-infected patients, we demonstrated that prophy angles used to clean and polish teeth can transmit HIV to cultures of human white blood cells, even after submerging the devices in 2 percent glutaraldehyde for two hours.[11] With prophy treatments, or cleanings, the boundaries between doctor and patient dissolve, as dentists have their own teeth cleaned by their hygienists. That's when visible amounts of the dentist's blood mix with the clear grease that's squeezed from a tube into prophy angles between patients.

The reason prophy grease has to be replenished is because it leaks out during use. But it's not a constant flow. Usually, it builds up until a dark glob of lubricant pops out as the hygienist is polishing a tooth, and then it gets scrubbed down into bleeding tissue with abrasive paste. The reason it's dark is because it's laden with blood. In the Florida dental practice where the HIV outbreak occurred, every patient that walked through the door was playing Russian roulette with Dr. Acer, but didn't know it. That is, not until Kimberly Bergalis would not rest until the state public health department, the CDC, or Congress became involved in finding out how and why she contracted HIV.

By experimenting with dental drills and prophy angles, I noticed visible amounts of blood occasionally coming out of the devices when I ran them in contact with a small amount of water in a porcelain container after they had been used in a bloody dental procedure. And that was after they had been cleaned and disinfected for use on the next patient. When I demonstrated this to my own dentist using prophy angles he had prepared for reuse, he looked down at the floor and said, "My God, I hope I haven't given my patients hepatitis." The next day, he instructed his staff to heat-sterilize his handpieces after every patient.

Research Published

Dr. Boe and I published our first paper in the *Journal of Clinical Microbiology* in 1991, demonstrating that dental handpieces could potentially retract and later eject an amount of blood from previous patients equivalent to a needlestick injury, which carries a risk of infecting one in three hundred patients with HIV.[12] We also showed that bacteria in blood could survive the superficial wiping with disinfectants, which is all that most dentists were doing with handpieces between patients. The following year, we submitted a second paper to the same journal, demonstrating that viruses also survived the high-level disinfection procedures recommended by the CDC and FDA. In that study, microbiologists at Loma Linda's School of Dentistry and the Retrovirus Clinic at Washington University's School of Medicine participated in the research. All three reviewers recommended that it be published, but the editor-in-chief rejected it.

He explained that our first paper was one of the most controversial the journal had ever published, and that this one would be even more controversial. I reformatted the paper and submitted it to *The Lancet* in London, while avoiding their editors in Washington, DC. I didn't want to take a chance on any connections other editors in the US may have had with editors at the

Journal of Clinical Microbiology. Lancet quickly accepted the paper with only a few minor grammatical changes.[13]

More HIV Cases

Compared with the HIV outbreak in Florida, potential problems with reusing unsterilized dental devices on HIV-infected patients were more clearly evident in the dental practice in Springfield, Massachusetts. In June 1989, James Sharpe, who was forty-three years old, made an appointment with Dr. Anthony Breglio. In contrast to Dr. Acer, Breglio was HIV-negative. Jim and his second wife, Jeanne, had recently moved back to Springfield from Miami, Florida. They worked together at the local Easy Quick Country Deli.

Jim hadn't seen a dentist for years, and was surprised to see how parts of the community where his dentist worked had deteriorated in recent years. He told me that the neighborhood around Breglio's office was riddled with crack houses, and a large house next to the dental practice had become a house of prostitution. Breglio, he said, even joked about working on the prostitutes. Jim had three teeth extracted, had several cavities filled, and was fitted with a partial denture. According to Jim, Dr. Breglio used a high-speed handpiece to section one of his molars.

Many dentists at the time would extend the burr—which is like a drill bit—then use it to section molars and extract them one piece at a time. This required burying the head of the handpiece in a profusely bleeding wound. This practice, which is not recommended, may be less common today, but it is still being done. About three weeks after having his teeth extracted, Jim developed night sweats and severe fatigue. He tested HIV-positive five months later, then developed AIDS in 1994.

Jim's physician concluded that his HIV infection was dental-related after an investigation by the state health department failed to discover any other probable cause. A second HIV-infected patient in Breglio's practice with no identified risk factors later surfaced after giving birth to an HIV-infected baby. Dental records showed that this patient had received dental treatments several weeks prior to Jim. Jim and his wife sued the dental practice.[14]

In the lawsuits filed by Kimberly Bergalis and others in Florida, and the Sharpes in Massachusetts, I was contacted to serve as an expert witness for defendants in the first case, and plaintiffs in the second. EPA approved of my testifying in my private capacity, provided that I donate my income from any expert witness fees to governmental or nonprofit organizations. Serving as an expert witness permitted me to have access to all of the patient records and other evidence in the cases, which were essential to my research.

The CDC never investigated Breglio's dental practice case as it did Acer's. Dr. Donald Marianos, head of the CDC's dental group, told me that investigating the Acer cases cost over $1 million. They simply didn't have the resources to take on another recall of thousands of patients. But the two practices were very similar insofar as the condition of their dental equipment. Dr. Breglio kept only a few old, high-speed handpieces, which he wrapped with gauze soaked in a dilute hypochlorite solution (Clorox) between patients, then later switched to another disinfectant. Dr. Breglio's handpieces, in fact, were so corroded that the metal was pitted with holes. To compensate for air escaping through the holes while he was drilling teeth, he increased the air pressure delivered by his compressors above normal.

BREGLIO ANSWERS SHARPE'S ATTORNEY (1994) [15]

In 1989, I used sodium hypochlorite to disinfect my handpieces. I noticed that excessive soaking of handpieces in sodium hypochlorite caused the handpiece motors (turbines) to rust and/or seize up.... I then substituted a different chemical, a phenol, to disinfect the handpieces.

Breglio's attorney was a heavyset, elderly gentleman with a very cantankerous courtroom demeanor. During the trial, I couldn't help but notice that he asked each of the plaintiffs' experts the same question to begin cross-examination: "Doctor, do you consider yourself to be a man of science?" The answer, of course, was always "Yes." And the attorney would continue to address the expert as a "man of science" throughout his cross-examination. His objective was to box the expert into a corner over some point of science, and then compel the witness to agree with the scientific literature on dental infection control published by professional and industry trade associations, such as the *Journal of the American Dental Association*. Then he always ended with the same question: "Doctor, do you have any proof that the AIDS virus was in the actual handpiece used on Mr. Sharpe?" The answer, of course, was always "No." Fortunately, I was the last expert to testify, which gave me plenty of time to think about my answers.

Before taking the stand, I noticed that the jury was mostly women. Then I looked at the back of the courtroom where all of the local and national reporters had lined up against the wall. All but one were women. Breglio's attorney began his cross-examination as expected: "Dr. Lewis, do you consider yourself to be a man of science?" I only got three words out—"Sir, your question..."—before the attorney hollered at the top of his voice: "*Just answer the question!*"

To which I replied with a Southern drawl, "Sir, your question is highly offensive to women in science. It doesn't matter whether I am a man or a woman. I would appreciate it if you would just call me a microbiologist."

The resulting effect exceeded my expectations. The attorney kept forgetting what I asked to be called, and he wasn't about to take any chance on calling me something different, which might offend some other group of jurors. When he came to the point in his cross-examination where he had addressed all the other experts as men of science, he had to stop and ask me to remind him of what I preferred to be called. Then, every time he paused to remember it, he would lose his train of thought.

He soon decided to wrap up his questions, and shoot off the one big question he always saved for last. He stood in front of me just barely long enough to ask, "Dr. Lewis, do you have any proof that the AIDS virus was in the actual handpiece used on Mr. Sharpe?" before turning his back to me and walking away.

"Yes," I said. He took several steps, then stopped and stood completely silent, still facing away from me. When he turned around, he spoke to me in a very non-lawyerly tone of voice. It was like we were two acquaintances having a conversation by the water cooler.

"What did you say?" he asked quietly.

"I said, 'Yes,'" I replied.

"You have proof that the AIDS virus was in the actual handpiece that was used on Mr. Sharpe?"

"Yes," I replied again.

"Is that proof in this courtroom?" he asked.

"Yes," I replied.

"Where?" he asked. I pointed to the stack of plaintiffs' exhibits resting on the table in front of him. He picked them up, brought them over to me, and said, "Show me." I sorted through them and picked out a random 8 x 10 color photo of a magnified view of one of Dr. Breglio's handpieces showing the head of the handpiece where the drill sticks out. Like all of the patient-ready handpieces he was using, the crevices around the O rings holding the drill were filled with blood and covered with transparent oil, which allowed the vivid bright red color to show through.

That blood, I explained, has collected from hundreds of patients or more. It's bright red because the lubricant keeps it from contacting the disinfectant, which would normally make it turn brown. In microbiology, I further explained, we apply *universal precautions* regarding the presence of HIV in blood samples. Absent any proof to the contrary, we must assume that it is present. This is particularly applicable when dealing with blood collected from hundreds of patients. So, I said, because I am a microbiologist, I must assume that this blood on the handpieces used on Mr. Sharpe did, in fact, contain HIV. The judge denied plaintiffs' requests to introduce evidence showing that Breglio's dental practice was located in a depressed area populated by drug addicts and prostitutes. Otherwise, I could have also pointed that out. According to the plaintiffs' evidence, Dr. Breglio had once boasted that a number of his patients were prostitutes. Breglio's attorney had no more questions. As I left the courtroom, a reporter from the *Boston Globe* came over and said she appreciated what I had to say about women in science.

In the end, the jury ruled that Breglio was negligent in failing to heat-sterilize his handpieces, even though the American Dental Association recommended soaking them in hypochlorite solution and other disinfectants. In my testimony, I provided documents from Midwest Dental Company, which manufactured the handpieces used by Dr. Breglio, and a copy of the ADA recommendations. Midwest recommended only heat-sterilization, and the ADA's recommendations stated that dentists should follow the manufacturers' recommendations.

Unfortunately, the jury sided with the defense's expert, John Molinari, on the question of causation. Molinari chaired the University of Detroit's Mercy School of Dentistry. He argued that flu-like symptoms Mr. Sharpe experienced several years prior to Mr. Sharpe's dental work may have been associated with HIV infection. These symptoms, however, were diagnosed and successfully treated as prostatitis.

Upon exposure to HIV, the virus proliferates in white blood cells throughout the body. Symptoms usually develop within two to four weeks, and include high fever (including night sweats), fatigue, headaches, and swollen lymph nodes. They typically last for only a couple of weeks.

Full-blown AIDS usually develops around five years later, and may take up to ten years or more.

From the beginning, I had urged Mr. Sharpe's attorney to let me and Mr. Sharpe's physicians explain the significance of Mr. Sharpe's severe night sweats and other symptoms consistent with initial HIV infection, called HIV viremia, which occurred three weeks after his extractions. Research has established that these symptoms help pinpoint the time of exposure to the AIDS virus.[16] Six months prior to Sharpe's trial, I wrote to his attorneys:[17]

LEWIS TO SHARPE'S ATTORNEYS (1995)

We need the medical records from Drs. Forgast and Villanueva giving any information they have documenting Mr. Sharpe's apparent HIV viremia (night sweats, fatigue) during July–December, 1989. (This is crucial to the case, in my opinion.) Also, we need their T-cell count documentation, showing the progression of Mr. Sharpe's HIV infection. If you have any of this information, please send a copy to me to cite in my affidavit.

Mr. Sharpe's attorneys, however, ignored my requests for this information. They wanted me to focus on what they believed to be the biggest challenge, which was proving that Dr. Breglio was negligent even though the CDC and ADA had recommended disinfection with hypochlorite and other germicides at the time Mr. Sharpe had his teeth extracted. I was far more worried about the defense lawyer confusing the whole issue of the incubation period for AIDS. Even at trial, I pleaded with Sharpe's lead attorney to let me rebut Molinari's testimony. He just told me not to worry.

Jurors were interviewed after they rendered the verdict:

BOSTON GLOBE QUESTIONS JURORS (1996)[18]

But, though the jury found Breglio's methods of sterilization inadequate, they were not convinced that the high-speed dental handpiece used in the procedure had transmitted the AIDS virus. "We all thought it was possible," said juror Gladys Sperry, 68, a retired musician from Belchertown. "But we had to decide whether it was probable." "We all felt there were too many other factors," Sperry said, "including the places he had lived, the length of time on the infection. You really didn't feel that it was more probable than not."

The verdict came after seven days of testimony, mostly from expert witnesses, about the probability of Sharpe's contracting the AIDS virus through the handpieces…. But the defense witness John Molinari…testified that it was "remotely possible" that the dental equipment was the cause of Sharpe's illness. He said low T-cell count suggested he had contracted the virus three or four years earlier."

I can't help but think about all of the resources it took to bring this case to trial, and how much it would have benefited public health to establish the first patient-to-patient transmission of HIV in dentistry. Instead, we ended up where we started, with many if not most dentists still wiping off dental handpieces and reusing them, while giving patients the "no-documented-cases" argument. In other words, so long as infections from dental drills aren't documented in the scientific literature, it's not a problem.

It reminds me of a conversation I had with a section chief at the FBI headquarters in Washington, DC, who called me a number of years ago to discuss some forensic evidence. When we started talking about jury trials, I commented on how I would hate for my life to rest on the ability of taxi cab drivers and plumbers to judge the reliability of the prosecution's arguments over DNA evidence. If the intent of the law is truly to arrive at the truth and achieve justice, why have a system where justice turns upon the ability of a retired musician from Belchertown to differentiate prostatitis from HIV viremia?

At least the sight of blood coming out of dental drills and prophy angles on *Primetime Live*, and in my photos that *JAMA* and others published, moved manufacturers to make handpieces heat-sterilizable, and back away from reusable prophy angles. Now, single-use prophy angles are much more common. That alone has made dentistry much safer. So, in the end, at least the manufacturers did their part—and without the government passing a single regulation!

In addition to the HIV cases in Florida and Massachusetts, I was also contacted by Bruce Williams, the father of Whitney Williams, an eleven-year-old girl living in Cook County, Illinois. Whitney, was one of ninety children in America with no known risk factors who were infected with HIV. Whitney's case related to my area of research because, at age two, she had four teeth extracted. Neither parent nor any of her four siblings were HIV positive.

Unfortunately, the Williams family began working with the Medical Legal Foundation in San Francisco to pursue a possible link between their daughter's HIV infection and oral polio vaccine, which was suspected of being contaminated with HIV-infected monkey cells. I say it was unfortunate because this action caused public health organizations to shift all of their attention toward protecting the polio vaccine and discrediting the parents. The US Department of Health and Human Services denounced any possible connection with the polio vaccine, and the Cook County Department of Public Health called upon the state attorney's office to review the case and determine whether Mr. Williams "had any contact with the gay community."[19] Any possibility of investigating the dental practice vaporized. Mr. Williams commented to the *Chicago Tribune*, "Why is this happening to us? We may not win the Good Housekeeping seal, but we are a good family."[20]

Sporadic Undocumented Cases

Re-lubricating dental handpieces and their attachments between patients, and soaking them with germicides, appears to have prevented widespread outbreaks such as the one that occurred in Florida. Sporadic infections involving only a few patients at a dental practice over the course of several years or more, however, are unlikely to ever be detected. Public health organizations lack the resources it takes to investigate such cases.

The total number of sporadic, undocumented cases could be large. According to the US Department of Labor, in 2010, there were approximately 84,000 general practice dentists and 182,000 hygienists actively working in the United States.[21] If each of them infected one patient with hepatitis every couple of years, that would be approximately one million sporadic cases falling through the cracks every 7.5 years. Sporadic dental infections, therefore, may play an important role in epidemics of blood-borne infections.

In response to the AIDS epidemic, the CDC advised dentists in 1986 to employ universal precautions and heat-sterilize handpieces and other reused devices whenever possible.[22] Because most handpieces could not withstand high temperatures, heat-sterilization of handpieces continued to remain low.[23] Dental infection control in other areas, however, did improve. An American Dental Association (ADA) survey in 1988 found that, over the previous two years, the use of gloves rose from 23 to 53 percent, needle-stick injuries decreased from 1.5 to 0.53 per 100 injections, and the number of dentists owning autoclaves increased from 67 to 80 percent.[24] Based on personal communications I had with manufacturers, the popularity of single-use prophy angles also began to rise at this time, and accounted for approximately 20 percent of the market by 1992. The ADA attributed much of the improvement to growing concerns that dentists were at elevated risk of HIV infection. Further improvements came in 1992 when our research prompted the FDA, CDC, and ADA to begin recommending only heat-sterilization for dental handpieces and their attachments.

National trends in acute cases of hepatitis B (HBV) and C (HCV) reported from 1986 to 1994 appear to reflect improvements in dental infection control, which began in 1988 (see Fig. 1 A, B). A national survey using blood samples collected from 1988 through 1994, however, failed to demonstrate any association between frequency of dental visits and the prevalence of infection.[25] Instead, decreases in reported cases of HBV, which is spread by sexual intercourse, dirty needles, and other routes involving direct blood-to-blood contact, have been primarily attributed to vaccination programs initiated in 1982.[26] But there's a problem with this explanation. In 1989 through 1995, a dramatic decline in HCV began at the same time HBV cases started to rapidly decline. HCV is primarily spread by blood-to-blood contact, not sexual intercourse; therefore, the vast majority of cases occur among injection-drug users sharing dirty needles. And, because HCV wasn't eliminated from the nation's blood donor supply until 1991 through 1994, and no vaccine is available, the reason acute cases began to steeply decline in 1988 through 1989 is not readily apparent.[27]

The current position taken by the CDC, the pharmaceutical industry, and most medical professionals concerning the hepatitis B and C epidemics in the United States leads to the following conclusions, which I consider to be questionable, at best:

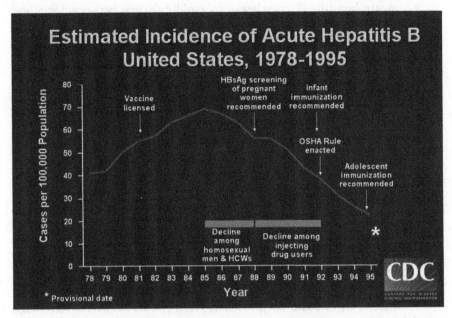

Figure 1A. Decrease in acute cases of hepatitis B (HBV) in the United States. Source: CDC.

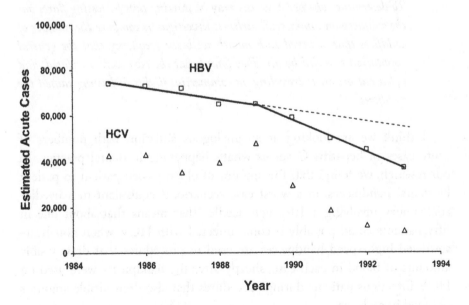

Figure 1B. Decrease in acute cases of hepatitis B (HBV) and C (HCV) in the United States. Source: David Lewis.

- Acute hepatitis B cases began to rapidly decline between 1988 and 1990 primarily because of vaccines that were first introduced in 1982.
- Because the abrupt decline in HBV cases was primarily caused by vaccines, and no vaccine is available for HCV, the causes behind the simultaneous abrupt declines in HBV and HCV cases are different and unrelated. In other words, it was purely coincidental that HBV and HCV cases both began to rapidly decline between 1988 and 1990.
- Because frequency of dental visits and prevalence of infection are unrelated, disrupting the exchange of visible amounts of blood via unsanitary dental devices had negligible impact on the spread of hepatitis C.

To illustrate what I think is missing in this picture, I give the following analogy:

Over a period of months, small groups of arsonists board commercial flights at random and toss flares out the window as they travel from one city to another. Eight board the first month, then twenty the second, twelve the next, seventeen the month after that, and one on the last the month.

To determine whether arsonists may be starting fires by tossing flares out the window on commercial airlines, investigators compare the number of wildfires that occurred each month with the frequency that the general population traveled by air. They found that the two were unrelated, and ruled out arsonists traveling on commercial flights as having caused the wildfires.

I think we are looking at an analogous situation with numbers of acute cases of hepatitis C versus what's happening in dental practice. In our research, we found that the amount of blood passed patient to patient by dental handpieces, in a worst-case scenario, is equivalent to a needlestick injury involving a 10-gauge needle. That means that about one in fifty patients could possibly become infected with HCV when non-heat-sterilized high-speed handpieces are used in procedures that draw visible amounts of blood in each case, shortly after the handpieces were used on HCV-infectious patients during procedures that also drew visible amounts of blood in each case.

Because most procedures involving high-speed handpieces don't involve visible amounts of blood, chances are small that any patient would be cross-infected with HCV. Again, we are talking about sporadic cases that go undetected because the mechanism is inefficient. High-speed handpieces wouldn't be expected to generate enough infections in a single dental practice to draw the attention of local public health officials.

That doesn't mean, however, that small numbers of low-risk patients in dental practices can't have a significant impact on the epidemic. Consider the impact that an HIV-infected airline steward could have had on the AIDS epidemic in its early stages. In every city where he had a layover, he could introduce the virus to the gay community, like tossing a flare into a field of dry grass. With HCV, I think it may be possible to sustain that kind of impact well beyond the genesis of the epidemic. It may happen whenever the high-risk population is relatively immobile compared with the low-risk population; the inefficient mechanism is widespread, and it occurs at a high frequency. Dental handpieces, for example, are used on thousands of patients day after day at every dental practice in every town and city across America. It doesn't take much imagination to envision an epidemic of bloodborne pathogens rapidly spreading coast to coast among isolated groups of inner-city injection-drug users if you have highly mobile dental patients sporadically acquiring the virus and spreading it to geographically isolated high-risk individuals.

Consider Jim Sharpe, for example. He was born in Enid, Oklahoma, moved to Guam, then to Honolulu, Hawaii, then Savannah, Georgia, then Plattsburgh, New York, then Springfield, Massachusetts, then Northampton, Massachusetts, then Miami, Florida, then Kansas City, Missouri, and on and on until he contracted HIV in a neighborhood populated with prostitutes and injection-drug users. If, instead, he had become chronically infected with HCV in Oklahoma, he could have passed it on to the injection-drug users in Massachusetts, and in other places along the way. Although he was not an injection-drug user himself, he often lived in neighborhoods with a prevalence of high-risk individuals, and potentially shared traces of blood with them whenever he had his teeth cleaned, filled, or extracted.

My point is that widespread inefficient mechanisms of transmitting infectious diseases may play a much larger role in epidemiology than is currently recognized. That's important to know. It means, for example, that the simultaneous abatement of the hepatitis B and C epidemics in the United States may not be coincidental, and may have had more to do with reducing sporadic bloodborne infections in dentistry than administering HBV vaccines, which certainly did nothing to prevent HCV infections.

To avoid scaring patients, the medical community tends to downplay the importance of inefficient modes of disease transmission that operate largely under the radar of our public health system. This reinforces widespread apathy within the profession. As memories of the Acer cases have faded, so has compliance with the CDC's guidelines recommending heat-sterilization of dental handpieces and other reusable dental devices. According to my contacts inside the industry, fewer than half of dentists operating in the Southern United States currently heat-sterilize their handpieces after every patient. Some estimate that less than 25 percent do this.

Muzzling the Messenger

Whenever government and industry fund universities to support government policies and industry practices, their first objective is to create a large body of supportive research in the peer-reviewed scientific literature. There appears to be no shortage of researchers at leading universities who are happy to take their money and publish whatever government agencies and corporations want. But what they publish isn't real science, and, therefore, they're not real scientists. It's all about marketing, and marketing, as a rule, involves deception. The "science" they create, whether intentionally or not, is just an illusion. Oftentimes the illusion is created simply by funding researchers who, as the old saying goes, are not the sharpest tools in the shed.

The second objective of government and industry is to protect that illusion, which often requires silencing scientists who disagree, especially whenever their research causes widespread public concerns over government policies and industry practices. Employers of scientists who step on the toes of government and industry are likely to be pressured by government agencies and industry trade groups to silence their employees.[28] As a result, employees may have their careers dead-ended, get fired, have their funding ended, be targeted with false allegations of research misconduct, or have any of a number of adverse actions taken against them. The severity of measures taken to silence scientists, at least in my experience, depends on the size of the toes being stepped upon. For example, I experienced little pushback from government programs and corporate interests involved in dental infection control. Government agencies and the wastewater treatment industry, by contrast, are pulling out all the stops to prevent me from talking about problems with treated sewage sludges, called biosolids. Every industrial and municipal polluter in the country has a huge stake in protecting lax EPA regulations, which allow pollutants concentrated at wastewater treatment plants to be applied to every available parcel of land, from farms and forests to school playgrounds.

In fact, during the entire time that I published research raising safety concerns over dental devices, no manufacturer in the United States or Europe ever once tried to interfere with my work. I came to know a number of the top executives of the leading manufacturers of dental handpieces at national trade shows, and toured their manufacturing operations. They never offered to fund my research, that is, "buy me out," and never hinted that I should do anything differently. All of our research was done on a shoestring, and I was able to pay for supplies and other expenses out of my own pocket.

FDA and CDC officials met and communicated with me from beginning to end. Both agencies were extremely supportive, even though I was highly critical of their policies. They never said or did anything to suppress our research or try to discredit me or any of my coauthors personally or professionally.

Toward the end of President Clinton's second term, he invited former FDA commissioner David Kessler to a Rose Garden ceremony when Congress voted to give the FDA new powers to regulate the tobacco industry. Dr. Kessler, at the time, was dean of the School of Medicine at Yale University. I happened to be catching a plane out of Reagan National later that afternoon. As I stood in line to have some changes made in my ticket, I looked back and noticed Dr. Kessler standing at a distance, staring at me.

When I was finished, he was still standing there watching me. Assuming that he was trying to place where he had seen me before, I walked over, stuck out my hand, and said, "Hi, I'm David Lewis."

He shook my hand, and said, "I know who you are. What are you working on now?"

We talked about flexible endoscopes for a few minutes, then I spent probably ten minutes or so talking with him about the tobacco industry. As we parted, he scribbled his address on a piece of paper so that I could drop him a line in the future.

The only flack I ever got from anyone over our research or criticisms of infection-control policies was from the ADA, and one of its member organizations, the California Dental Association (CDA). The president of the CDA wrote a letter to my EPA laboratory director complaining about my research. My director responded that EPA had no problem with my outside activities, and EPA headquarters later approved my dental infection control research as part of my official EPA duties. The ADA published an editorial claiming that I was a dentist who stood to profit from a patent on a disinfection procedure if heat-sterilization became the new standard of practice. I've never been a dentist, owned a patent on a dental product, or made any money from selling dental products of any kind.

The ADA's president became a vocal critic of me, personally. One of my coauthors was a faculty member at a dental school where the ADA president gave the commencement address. My coauthor said that a number of faculty members boycotted the event because of the ADA president's personal attacks. In the end, however, the ADA invited me to meet with CDC, FDA, and ADA officials at their headquarters to discuss the wording of the ADA's new guidelines recommending heat-sterilization of dental handpieces.

ASK YOUR ENDOSCOPY CLINIC

DO YOU STERILIZE WITH PERACTETIC ACID?

By age fifty, everyone should be routinely screened for colorectal cancer. I am, and fortunately have been cancer free so far. Colon cancer is one of the easiest cancers to cure *when it's caught in time.* Flexible fiber optic endoscopes, which are long flexible tubes with a camera lens at the end, are technological marvels. It's pretty amazing how doctors today can routinely inspect internal areas of the body such as the colon and lungs, collect biopsies, and perform corrective surgery. The only problem is that flexible endoscopes cannot withstand heat-sterilization, and most models aren't infection-control friendly. For example, internal air and water channels in most flexible endoscopes are too small to insert brushes all the way through them to clean out blood, feces, and tissue that collect inside.

Pentax flexible endoscope with all channels fully accessible to brushing. Photo courtesy of Walker Montgomery.

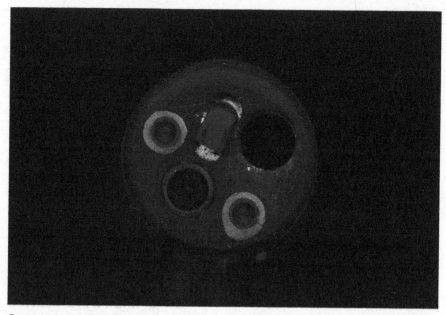

Insertion tip of an Olympus colonoscope with lubricants and live bacteria (fluorescing under U.V. light) leaking from a non-brushable air-water channel after preparing the scope for the next patient according to the CDC's current disinfection guidelines. Courtesy David Lewis.

Tips on Choosing the Best Endoscopy Clinic[1]

1. **Sterilization is better than disinfection.** Although most doctors assure patients that the flexible endoscope used on them will be *sterilized*, most endoscopy units (approximately 80 percent) only *disinfect* them. Disinfection doesn't kill spores, which are produced by some very important pathogens such as *Clostridium difficile* ("C diff") and *Mycobacterium*. In 2012, the CDC estimated that C diff, which causes an increasingly fatal form of severe diarrhea, is linked to fourteen thousand American deaths each year.[2] Currently, Steris Corporation's process using buffered peracetic acid is the only FDA-approved process in common use for sterilizing flexible endoscopes. It is used in approximately 20 percent of the endoscopy clinics in the United States.[3]

2. **Flushing is better than soaking.** Liquid chemical germicides used to disinfect or sterilize flexible endoscopes are *mass-transport limited*. That means disinfection and sterilization rates increase proportionally to the solution flow rates. In other words, doubling the flow rate of a disinfectant solution doubles the rate of disinfection. That's critically

important with flexible endoscopes because their internal channels are contaminated with biofilms, and particles of blood, feces, and other patient materials hide in nooks and crannies. Doubling the flow rate doubles the chances that the disinfectant will penetrate this type of contamination. While many endoscopy clinics use machines that automatically circulate germicide solutions over the internal and external surfaces of flexible endoscopes, most are only soaked for short periods of time (approximately ten minutes).

3. **Sporadic infections are underreported.** Your doctor may reassure you that only one patient is infected per 1.8 million procedures using flexible endoscopes. But bear in mind that this information is based on surveys of practitioners and reported outbreaks, where numerous patients are infected at or about the same time at the same facility. Outbreaks are relatively easy to detect and report compared with sporadic cases involving a single patient; also, outbreaks are caused by egregious lapses in infection control. Most infections associated with flexible endoscopes, by far, are sporadic. Patients are unlikely to suspect endoscopes as the culprit, especially when doctors falsely reassure patients that endoscopes are sterilized when most (80 percent) have only been treated with disinfectants. Most flexible endoscopes are only soaked in disinfectant for ten minutes, and it would take hours to reach sterilization conditions. And, if you test positive for hepatitis C, anal HPV infection, or other infectious agents transmitted by flexible endoscopes, the last thing any doctor will ask is "Have you had a colonoscopy done?"

4. **Don't avoid endoscopy.** The risks associated with ignoring colorectal cancer and other serious diseases far outweigh the risks of infection whenever CDC guidelines are followed for disinfecting flexible endoscopes. Just be smart about it when it comes to choosing where you have endoscopy done and what kind of disinfection or sterilization process is used. If you still have concerns, you should talk with your doctor about alternatives, such as radiology.

Disinfection versus Sterilization

Infection-control guidelines are based on E. H. Spaulding's classification system, which recommends high-level disinfection as the minimum standard for flexible endoscopes. Spaulding considered endoscopes to be *semi-critical* devices, which enter unsterile areas of the body and contact only intact mucous membranes.[4] Some bleeding, however, occurs in half or more of all colonoscopies, for example, when biopsies are taken. Also, the inner lining of the intestine is easily damaged when flexible endoscopes turn corners, press against hemorrhoids, and contact weakened or injured tissues. Currently, buffered peracetic acid and ethylene oxide gas (ETO) are the only FDA-approved sterilization processes available for flexible endoscopes.[5]

The Spaulding classification system provided a rationale for not sterilizing flexible endoscopes at a time when ethylene oxide sterilization, an overnight process, was the only sterilization process available for flexible endoscopes. It gave rise to some irrational distinctions. For example, biopsy forceps must be sterilized, while the biopsy channels through which the forceps are inserted in flexible endoscopes require only high-level disinfection. This makes little, if any, sense when biopsy channels are contaminated with the same patient materials as the forceps during endoscopic procedures.

Most physicians still think it's unnecessary to sterilize a device that's inserted in unsterile areas of the body, especially the colon where large numbers of fecal coliform bacteria proliferate. The presence of fecal matter, however, offers little if any protection from hepatitis B, HIV, and other pathogens that contaminate flexible endoscopes. Otherwise, sexually transmitted diseases among homosexual men would be less of a public health concern.

Also, when the Spaulding classification system was published in 1968, the ease by which some viruses can pass through intact mucous membranes was unknown. Now, it is widely recognized that viral transmission across mucous membranes, such as HIV transmission via breast-feeding, is an efficient mechanism of disease transmission.

More importantly, few individuals with severely compromised immune systems survived long enough to be treated with endoscopes when the Spaulding system was put into practice. Now, organ transplant recipients, patients undergoing chemotherapy, diabetics, AIDS patients, and people with a host of other conditions that compromise their immune systems are surviving for many years and receiving frequent endoscopic examinations. Introducing new strains of organisms into immune-compromised patients can lead to severe disease or death. Therefore, it is no longer safe to assume that patients can fend off low numbers of opportunistic pathogens known to escape high-level disinfection. Moreover, because resistance to antibiotics has become a widespread problem, many endoscope-related infections that were once easily treated are now life-threatening.

Although the amount of infectious matter potentially passed patient to patient by flexible endoscopes is small, we found that even a few microliters of HIV-positive blood contaminating lubricants used in endoscopes and dental handpieces can efficiently transmit the virus to human lymphocytes.[6] Admittedly, the risk of HIV infection from endoscopes is very low. Endoscopic procedures, however, may be a far more common, yet largely undocumented, source of infection with other viruses, such as human papillomavirus and cytomegalovirus.

Hidden Problems

In the United States, Europe, and elsewhere throughout the world, current guidelines for reprocessing flexible endoscopes recommend pre-cleaning followed by high-level disinfection.[7] The adequacy of these guidelines, according to their authors and proponents, rests on the efficacy of high-level disinfectants in laboratory tests and the low numbers of documented cases of infection. Experimental designs of the tests cited, however, bear little resemblance to the practice of endoscopy, where pathogens lodge in difficult-to-clean areas in bits of flesh, blood, and feces hardened with glutaraldehyde and mixed with viscous lubricants. Moreover, a patient infected by an endoscope has almost no chance of having his case documented in the peer-reviewed medical literature.

Endoscope manufacturers admit that visible traces of blood, feces, and other patient materials remain in internal areas of flexible endoscopes, serving as a potential source of infection despite our best efforts to clean the devices.[8] Even a cursory application of the fundamental laws of physics reveals the challenge this presents to infection control. It would take at least fifty-eight hours for 2 percent glutaraldehyde, the germicide most commonly used on endoscopes, to diffuse through even barely visible traces of patient material.[9] For germicides to reach pathogens buried in such material in less than an hour, the concentration and temperature of the germicide would have to be so high that it would quickly destroy endoscopes and be far too hazardous to handle. Microbiologists have long recognized this limitation and stressed that all surfaces must be thoroughly cleaned for disinfection to work properly.

What has not been appreciated, however, are the pitfalls of testing chemical germicides using standard techniques for culturing microbes.

When germicides contact residual patient debris inside endoscopes, they kill microorganisms only on surfaces of the debris. This can leave large numbers of viable organisms hidden inside. Whenever such superficially disinfected material is submerged in nutrient solutions to test for live organisms, nutrients fail to penetrate the material. Few, if any, colonies of microorganisms are revealed. What appears to be adequately disinfected material can actually harbor large numbers of pathogens buried inside, waiting to be dislodged from air/water and biopsy channels during endoscopic procedures. Once the material enters patients and is broken down, these stowaway microorganisms can begin multiplying and cause infection.

Lubricants used on external surfaces and mechanical parts of endoscopes add yet another dimension to the problem. Even the most fragile viruses, including human immunodeficiency virus (HIV), can survive a two-hour exposure to 2 percent glutaraldehyde when entrapped in lubricants.[10] Once again, large numbers of microorganisms can go undetected by conventional microbiological tests.[11]

Poor Compliance

Only about one in ten endoscopy units follow FDA requirements for properly disinfecting and sterilizing endoscopes.[12] These requirements are supported by endoscope and disinfectant manufacturers, and no infections have been reported when complying with these requirements. Most health-care facilities (84 percent) currently follow less stringent recommendations of the Centers for Disease Control and Prevention and professional endoscopy societies. These recommendations are inadequate for achieving the minimum conditions of exposure time and temperature required to kill *Mycobacterium tuberculosis* and other germicide-resistant pathogens with 2 percent glutaraldehyde. Outbreaks involving *Mycobacterium* and other pathogens have occurred when applying these less stringent guidelines.

One national survey found that most physicians (66 percent) reprocess flexible endoscopes in considerably less time than is required to perform the cleaning and disinfection steps recommended in infection control guidelines.[13] While every effort should be made to improve compliance, appropriate recommendations should provide the widest margin of safety practical to protect patients when the performance of reprocessing procedures is less than perfect. With this in mind, the following general approach to reprocessing endoscopes provides the best margin of safety in current practice.

Flexible endoscopes should be meticulously cleaned according to manufacturers' instructions as soon as they are removed from patients, and then subjected to a sterilization procedure. All germicides approved by the FDA for use on flexible endoscopes are chemical sterilants; that is, they have the *potential* to achieve sterilization, given sufficient exposure times. For example, the most common liquid chemical germicide used, 2 percent glutaraldehyde, can achieve sterilization with a minimum of ten hours of exposure time. But the maximum exposure time used in practice is only forty-five minutes, and most facilities use only ten minutes of exposure.

Another important point is that liquid chemical germicides that facilitate rather than frustrate the cleaning process should be chosen. Two of the most common germicides used on endoscopes, glutaraldehyde and peracetic acid, are used for other purposes that illustrate an important difference between these two chemicals. Like its first cousin formaldehyde, glutaraldehyde is used to prevent decomposition. To preserve a frog so that it can be dissected and studied fifty or one hundred years later, just treat it with glutaraldehyde. But, when scientists want to *dissolve* organic matter instead of preserve it, peracetic acid is used. For years, researchers have employed this powerful oxidizing agent to render inaccessible areas of intricate laboratory glassware completely clean of even baked-on organic matter.

The ability of buffered peracetic acid to remove glutaraldehyde-hardened patient material from biopsy channels has been demonstrated using surface infrared spectroscopy.[14] What has long been common knowledge to laboratory researchers wanting to either preserve organic material with glutaraldehyde or remove it with peracetic acid can be easily visualized by even the most casual observer.

Some health-care practitioners switching from soaking endoscopes in glutaraldehyde to purging the devices with buffered peracetic acid have experienced an initial increase in repairs. The oxidizing power of peracetic acid causes it to completely break down proteins and other organic matter. As hardened patient material begins to dissolve away, leaks are exposed in internal channels. Incredibly (or disgustingly), many flexible endoscopes now in use apparently are literally held together by potentially infectious patient material, sealed by a "superglue" containing microorganisms buried between layers of glutaraldehyde-hardened surfaces.

The oxidizing action of peracetic acid is short-lived. Just as hydrogen peroxide decomposes into innocuous byproducts (water and oxygen), peracetic acid quickly breaks down to oxygen and dilute acetic acid—a weak vinegar solution. Peracetic acid in the dilute buffered solution used for sterilization does not require fume hoods and can be disposed of down the drain after the solution exits the sterilization unit. Glutaraldehyde, on the other hand, is persistent, and breathing its vapors tends to preserve human lungs, like frogs, only not in working condition.

Sheath Technology

Another method of rendering flexible endoscopes safer for reuse is newly developed sheath technology.[15] Rather than trying to clean contaminated surfaces, they are simply discarded after each use of the flexible endoscope. Discarded portions include all external surfaces that directly contact the patient as well as internal channels that can retain bodily fluids and other patient matter. Optical and mechanical workings of sheathed endoscopes are designed as a separate unit that is re-sheathed and supplied with sterile, disposable biopsy and air/water channels after each use. Although Vision-Sciences, Inc., developed the basic technology, a number of other endoscope manufacturers, including Olympus Optical Co., Ltd., have considered sheathed endoscopes.[16]

No Documented Cases

Proponents of high-level disinfection of endoscopes point out that only one infection has been reported in the medical literature for an estimated 1.8 million endoscopic procedures.[17] In most cases, failure to follow all of the recommended steps for pre-cleaning and high-level disinfection is blamed for causing these infections. These numbers are meaningless, however, considering that the likelihood that any endoscope-induced infection will be published in the medical literature is probably also on the order of one in a million.

When proponents of disinfection refer to reported cases, they often fail to point out that a number of infections have been documented where adherence to guidelines for pre-cleaning and high-level disinfection were followed. By comparison, no infections have been reported with proper pre-cleaning and sterilization. Sterilization, microbiologists agree, offers a higher degree of assurance that microorganisms are killed. *Mycobacterium* passed patient to patient via bronchoscopes in a Roanoke, Virginia, hospital, for example, could not be eliminated by even the most rigorous pre-cleaning followed by high-level disinfection with 2 percent glutaraldehyde, but was eliminated with peracetic acid sterilization.[18] Even manufacturers of high-level disinfectants acknowledge that bacteria have demonstrated an increasing resistance to glutaraldehyde while remaining susceptible to peracetic acid.[19]

A 1992 survey found that 6 percent of US nurses were knowledgeable of outbreaks of infections attributed to flexible endoscopes at their institutions.[20] Recent advances in DNA fingerprinting enabled several cases of tuberculosis transmission via endoscopes treated with 2 percent glutaraldehyde to be detected in separate hospitals.[21] One patient died from the nosocomial transmission. If this new science of molecular epidemiology were to be widely applied in the area of endoscopy, I expect that the numbers of infections uncovered and scope of organisms involved would quickly precipitate an upgrading of federal guidelines to a sterilization standard.

In my own work, I have encountered scores of unreported cases where patients with no other identified risk factors have developed life-threatening infections within expected incubation times after both dental and endoscopic procedures.[22] For example, I have investigated several cases where male patients developed human papillomavirus (HPV) infections of the anus, which began to appear four to five weeks after routine colorectal examinations in separate medical facilities. The flexible endoscopes used on these patients had been cleaned and subjected to high-level disinfection with 2 percent glutaraldehyde. Most strains (types 6, 11, 16, 18) of HPV, the virus causing genital warts, place patients at high risk of developing anal cancer when the virus infects this area of the body. Ironically, these patients submitted to endoscopic examinations after being encouraged by their physicians to have annual screening tests for colorectal cancer.

Patient Concerns

While patients should not be frightened away from needed medical care, neither should they be kept uninformed. Physicians should encourage candidates for endoscopy to discuss any concerns they may have about infection control and address their questions fully and honestly. Just as physicians tell patients that paralysis from surgery or death from anesthesia is a remote but real possibility, they should make patients aware of the risks of infection from endoscopy. Until sterilization becomes the standard of practice, patients should be fully informed and afforded the opportunity to decide whether to disregard the small chance of infection from unsterilized endoscopes or seek the added assurance of safety provided by sterilization.

Recommended guidelines for reprocessing endoscopes should be upgraded from high-level disinfection to the exclusive employment of sterilization methods and sterile disposable products. The federal Centers for Disease Control and Prevention has already instituted this change in dentistry with regard to dental handpieces and similar devices retaining blood and other patient materials in difficult-to-clean areas.[23] There, the standard is simple and can be understood by any patient. Any device entering the oral cavity must be either subjected to an approved sterilization procedure after each use or discarded.

There is no reason why government agencies and medical associations should not adopt this same high standard for endoscopy in response to the rapidly changing world of medical technology and increasing microbial resistance to high-level disinfectants. Sterilization is the standard of care that most physicians already tell their patients they are receiving for endoscopy, and the majority of patients mistakenly believe this is the standard of care they are receiving.[24]

Muzzling the Messenger

As was the case with my research on dental infection control, the CDC and FDA were both supportive of my efforts to improve infection control standards regarding endoscopy. My research published in *Nature Medicine* demonstrated that lubricants used with flexible endoscopes could thwart disinfection with glutaraldehyde, the most common germicide used by endoscopy clinics.[25] A cover story about my research on disinfectants by *Hippocrates* magazine (Time Inc.) won a national award, and lengthy articles focusing on problems with using glutaraldehyde to disinfect flexible endoscopes were published by *USA Today, Newsweek,* and others.[26] Also, network television news programs, including Dr. Timothy Johnson with ABC's *Good Morning America*, NBC's *Nightly News*, and PBS's *Healthweek*, interviewed me on the subject.[27] Thus, the threshold for silencing me—causing widespread public concerns—was passed.

Nevertheless, the medical community responded appropriately. For example, *Practical Gastroenterology*, a peer-reviewed medical journal, devoted two issues to covering the debate by inviting me to contribute an article, and others to argue that it isn't necessary to sterilize flexible endoscopes.[28] Only one company, Custom Ultrasonics, Inc., which manufactures a washer-disinfector that uses glutaraldehyde, retaliated. After the *Los Angeles Times* refused to publish the company's false allegations of research and ethics misconduct against me, its infection-control chief, Lawrence Muscarella, published them in the company's newsletter.[29] Specifically, the company alleged that I failed to disclose a "financial relationship" with Steris Corporation when I published our paper in *Nature Medicine*.

As editors of *Hippocrates* magazine pointed out, Steris Corporation had once donated funds to a community church that David Gattie and others, including myself, built in Oconee County, Georgia, in 1993. The church, called Saxon Road Church, held regular Sunday services until 2013, when David retired as pastor. It had an all-volunteer staff, and didn't solicit tithes or offerings. Custom Ultrasonics refused to publish a brief response I submitted regarding its allegations. In it, I stated that I did, in fact, disclose the donations, which were approved by EPA ethics officials. But, because I could not financially benefit from the church, editors at *Nature Medicine* didn't consider the donations to represent a financial conflict of interest. All I can do is disclose any possible conflicts of interest. It's up to the editors whether they choose to publish them.

3

ASK YOUR GARDEN CENTER

DOES THIS MULCH CONTAIN
BIOSOLIDS?

In 1978, a soil scientist working in EPA's Office of Water (OW) by the name of John Walker advised deputy assistant administrator Henry Longest, "The application of some low levels of toxic substances to land for food crop production should not be prohibited." Walker had transferred to EPA several years earlier from a USDA laboratory run by Rufus Chaney, an agronomist studying land application of treated sewage sludges, called biosolids.

And so began what I would argue is the worst conceivable plan EPA could have ever come up with for containing and disposing of the nation's industrial and municipal pollutants. That plan was to discharge "pretreated" hazardous industrial wastes into sewer systems, treat the sewage sludges that settle out at wastewater treatment plants with chemical and biological processes to reduce pathogens and odors, and then market the treated sewage sludges as cheap fertilizer products.

To appeal to consumers, EPA and the wastewater industry refer to land application of biosolids containing industrial chemicals as "recycling."[1] But, as the international symbol for recycling illustrates, recycling is a cycle. To reduce wastes, products made of metal, glass, and plastic are recycled to make other metal, glass, and plastic products. They go back to manufacturing plants where they started, and are remade into more of the same kinds of products. Cycles are perpetual—you keep using the material over and over.

So, if you really want to recycle industrial chemicals in biosolids, take your biosolids back to Monsanto, Dow Chemical, Merck, and Ciba-Geigy and ask them to make some more pharmaceuticals and pesticides out of it.

EPA does not require that composted sewage sludges containing pharmaceuticals and other industrial pollutants be labeled. Hence, companies selling mulch products omit this information on their websites and product labels. They entice customers with deceptively named products, such as Earthfood, Meadow Life, and Nutra-Green. They claim that biosolids are *natural organic* products; unlike synthetic fertilizers, they *purify* the soil with millions of microorganisms and produce more nutritious vegetables.

There's nothing natural, however, about Prozac and other synthetic chemical pollutants in composted sewage sludge. And the type of aerobic bacteria naturally found in garden soils would be better off without loading them up with anaerobic bacteria from sewage sludges. In a study published in *Nature*, we found that, unlike commercial inorganic fertilizers, sewage sludges could increase the persistence and toxicity of the degradation products formed when pesticides, pharmaceuticals, and other pollutants break down in soil.[2]

Environmental scientists are concerned about synthetic chemicals, such as Prozac, because nature has not had much time to adapt to them. Nitrogen in the form of ammonium and potassium nitrates found in commercial fertilizers, however, has been part of nature as long as proteins and other forms of organic nitrogen. It makes no sense to market composted sewage sludge on the basis that the kind of inorganic nitrogen found in commercial fertilizers isn't natural. And, it's deceitful to sell composted sewage sludge—probably the richest source of complex manmade chemicals on Earth—on the basis that it's *natural*.

At the 2012 annual meeting of the Society of Environmental Toxicology and Chemistry in Long Beach, California, Professor Rebecca Klaper presented her research on traces of Prozac in effluent from wastewater treatment plants. She found that as little as one part per billion scrambled, as environmental reporter Brian Bienkowski put it, how genes in the brains of developing fish turn on and off.[3]

Reproduction rates fell, and males, which became more aggressive, killed some of the females. Klaper, according to Bienkowski, said, "There appeared to be architectural changes to the young minnows' brains."[4] Bienkowski quoted a chemist at Los Angeles' wastewater treatment plant, who commented that Tegretol, another antidepressant in Klaper's study, enters the plant and comes out at nearly constant levels, meaning that the city's treatment processes have no effect on it.

Tegretol is one of many neurotoxic chemicals found at ppm-levels in most sewage sludges. In 2009, EPA published its results from eighty-four samples of sewage sludge collected from seventy-four wastewater treatment plants across the county. It found Prozac (fluoxetine) in seventy-nine of the samples, which ranged from 12.4 to 3,130 parts per billion. That's twelve to over three thousand times higher than the minimum concentrations that caused the brain damage observed in Klaper's study. Like many drugs, Prozac is soluble in body fat. So, it tends to concentrate in sewage sludges, which contain a lot of animal fat. Chemicals increase in neurotoxicity with increases in fat solubility.[5] Sewage treatment plants, therefore, concentrate the most neurotoxic pollutants in sewage sludges.

Most wastewater treatment plants just add lime to their sewage sludges to reduce odors and pathogens, and give it away for application to farms, forests, school playgrounds, and other public and private lands. I mention playgrounds because, in 2007, Milwaukee Public Schools closed thirty fields and playgrounds after the city discovered that its "Class A" biosolids were highly contaminated with PCBs. Until the topsoil could be removed and buried at a hazardous waste site, the city fenced off fields where over sixteen thousand youths and adults played softball, soccer, and kickball. Milwaukee is one of the few cities that monitors its biosolids, which it calls *Milorganite*, for PCBs.

In 2002, a research group I assembled at the UGA documented the first cases of biosolids-related illnesses and deaths in the peer-reviewed scientific literature.[6] We evaluated both affected and unaffected residents living within approximately one-half mile (1 km) of land application sites. Most complained of irritation (e.g., skin rashes and burning of the eyes, throat, and lungs) after exposure to winds blowing from treated fields. Approximately one-fourth of fifty-four individuals exhibited *Staphylococcus aureus* infections of the skin and respiratory tract, including two mortalities from septicemia and staphylococcal pneumonia.

Below are a few excerpts from a speech by Joanne Marshall, whose son developed difficulty breathing and died. He and most residents in their neighborhood in Greenland, NH experienced burning lungs and difficulty breathing as soon as an endless procession of trucks began dumping biosolids near their homes.[7]

As days and weeks went by we became sicker and sicker. Not just my family, but our whole immediate neighborhood. We all shared the same symptoms; first nausea and vomiting followed by severe stomach cramps and migraine headaches. Then fever and flu-like symptoms, more respiratory. There was a continual battle of thick mucous, one that made it hard to swallow and discharge. At times it appeared as if our reflex system had slowed because you would gag on the mucous and sometimes choke to dispel it. Often it would wake you because your breathing passages were blocked by it.

[On] the evening of Thanksgiving, I kissed my son, Shayne of 26 years, goodnight for the last time. Around four a.m. that morning, I was awakened to a frightful scream from my other son, who was home from college during the holiday. When I ran to the room, Shayne appeared unconscious, yet he seemed like he was gasping. 911 was called and all I could do was hug him and wait for the paramedics. We spent what seemed like an eternity in the hospital waiting room, only to be told my son was dead.

Nature applauded our work in an editorial and news article, calling EPA's biosolids program *an institutional failure spanning more than three decades—* and the *presidential administrations of both parties.*[8] It cited a 2007 study confirming the link we reported between exposures to biosolids and gastro-intestinal and respiratory symptoms.[9] Symptoms included excessive secretion of tears, abdominal bloating, jaundice, skin ulcers, dehydration, weight loss, and general weakness. Reports of bronchitis, upper respiratory infection, and giardiasis were also statistically significantly elevated.

Thus, EPA encourages land application of biosolids, which contain concentrated (ppm and higher) levels of the same pollutants that the Clean Water and Clean Air Acts prohibit industry from dumping in our waterways and spewing from smoke stacks. The reason is simple. There is no Clean Soil Act. Some hazardous wastes are *pretreated* before they are discharged. Most hazardous wastes, however, don't have to be pretreated, and pretreatment can create many more problems than it solves. That's because pretreating hazardous wastes with chemical and biological processes transforms them into even more complex mixtures of new pollutants for which we have no analytical methods and no scientific data regarding their effects.

What scientists do know about the degradation products of pollutants, regardless of whether they originate from pretreatment or natural environmental processes, is that they can be as harmful, if not more harmful, than the parent compounds.[10] The only way to make complex organic chemical pollutants become non-carcinogenic, non-mutagenic, and nontoxic is to use extreme heat to break them down to their simplest building blocks, that is, CO_2 and water, and then recycle or immobilize any residual heavy metals and radionuclides.[11]

Far-Reaching Implications

When problems with EPA's biosolids program have surfaced in the past, EPA called it "the little program that gets a lot of attention." This program is anything but little, and it needs a whole lot more attention. It embodies the reason EPA was created, and forms the underlying premise for the agency's overall approach to environmental protection. So what was EPA going to do with all of the pollutants that Congress decided should be removed from our air and water? Put them in our soil, of course. Of the four classical elements of nature—earth, water, air, and fire—Earth was the only place Mother Nature had left where cities and factories could still dispose of their chemical and biological wastes cheaply. But choosing the cheapest option in the beginning often turns out to be the most costly in the long run. A much better choice would have been to use fire to break pollutants back down to their most basic elements.

The toll EPA's approach may have on public health and the environment in the long run is immeasurable. Different genetic variants of plants, animals, and microorganisms vary widely as to the kinds of chemical pollutants they can tolerate in air, water, soil, and food. Spreading mixtures of all of the chemical pollutants produced in the city Los Angeles, for example, on land—thus contaminating the soil, air, and water with virtually every known and unknown environmental pollutant—runs contrary to common sense.

The chances that any one species will be seriously harmed by environmental traces of any one of the chemical pollutants produced by modern industrial societies may be small. But the chances that serious adverse effects will occur from exposing any living organism to a complex and unpredictable mixture of tens of thousands of chemical pollutants at parts-per-million (ppm) levels is likely a virtual certainty. This is particularly true for exposures that occur during developmental stages, and over multiple generations.

Endocrine-disrupting chemicals, for example, can cause cancer, birth defects, and other developmental problems at parts-per-trillion (ppt) levels.[12]

Wastewater treatment plants are particularly problematic because they magnify the concentrations of fat-soluble pollutants, which have the highest potential for causing adverse health effects over time. Just compare the concentrations of fat-soluble carcinogens, endocrine disruptors, and neurotoxins in private wells on farms, for example, with the concentrations of these same chemicals in sewage sludges at any wastewater treatment plant.[13] Industrial and municipal pollutant concentrations are oftentimes a million-fold higher in biosolids compared with their levels in drinking water (parts-per-million versus parts-per-trillion).

To be exposed to biosolids on a regular basis, you don't have to live in rural areas where tons per acre are regularly applied for agricultural use. Every city and town is connected to a wastewater treatment plant, and over half of America's sewage sludge is land-applied. To get rid of their sewage sludges, cities spread them on every available spot of land, from public parks and school playgrounds to golf courses and the medians of interstate highways. Sewage sludge is sprayed on forests and sold in composted mulch at local nurseries. EPA even promotes spreading it on poor, inner-city neighborhoods to reduce the risks of lead poisoning in children who eat it. And, if you don't live near a big city with lots of industrial wastes being discharged into sewers, you still have cause for concern. Some of the composts you buy at your local garden center, such as Milwaukee's product, Milorganite, can contain complex mixtures of pharmaceuticals and other chemical wastes.

Unless you believe in magic, there's nothing safe about spreading pollutants on land in concentrations higher than what is known to be unsafe in air and water. When someone drinks water from a private well contaminated with biosolids applied to surrounding land, or inhales air blowing from a school playground treated with biosolids, it doesn't matter whether the chromium VI and endocrine disrupters originated with the air or water, or the biosolids that contaminated them.

The percentage of sewage sludge that is spread on land, compared to what is incinerated or landfilled, is steadily increasing. That's because EPA continues to fund the Water Environment Federation, the North East Biosolids and Residuals Association (NEBRA), and other lobbying and industry trade organizations to aggressively promote the "recycling" of biosolids as safe and environmentally beneficial. EPA and some of these organizations it funds appear to intentionally mislead the public concerning what's actually in biosolids, and the risks it poses (see Appendix I: Ten Myths About Biosolids). NEBRA, for example, states on its website:[14]

> *Pretreatment regulations were developed under the US Clean Water Act. Similar requirements are applied in some Canadian provinces. These regulations ban the discharge of any toxic substance that might…*
>
> • *hinder the wastewater treatment process*
> • *pass through the wastewater treatment plant and contaminate the plant's receiving waters, or*
> • *concentrate in the biosolids and affect the ability to recycle them.*

Contrary to what NEBRA is telling the public, even EPA published a long list of pollutants found in most, if not all, sewage sludges, which are known to cause serious adverse health effects. In its Targeted National Sewage Sludge Survey (TNSSS) of seventy-four wastewater treatment plants across the United States, EPA, for example, reported the following pollutants at concentrations reaching parts-per-million (ppm) levels:[15]

Heavy metals, e.g., chromium (6.7–1,160 ppm), nickel (7.5–526 ppm), lead (5.8–450 ppm), molybdenum (2.5–132 ppm), mercury (0.2–8.3 ppm)

Polycyclic aromatic hydrocarbons (PAHs) and semi-volatiles, e.g., bis (2-Ethylhexyl) phthalate (0.7–31 ppm), Benzo(a)pyrene (0.06–4.5 ppm)

Polybrominated diphenyl ether congeners (PBDEs), e.g., BDE-209 (0.15–17 ppm), BDE-47 (0.07–5 ppm), BDE-99 (0.06–4 ppm)

Pharmaceuticals, e.g., ciprofloxacin (Cipro) (0.07–48 ppm), carbamazepine (Tegretol, Equetro) (0.01–6 ppm), fluoxetine (Prozac) (0.01-3 ppm)

Steroids and hormones, e.g., androsterone (0.02–1 ppm), estrone (0.03–1 ppm)

Environmentally triggered neurological and autoimmune diseases and disorders are rapidly increasing in prevalence in industrialized areas of the world. This includes, for example, autism spectrum disorders, Alzheimer's disease, Parkinson's disease, diabetes, and rheumatoid arthritis. Scientists generally agree that traces of environmental pollutants play an important, if not the most important, role in triggering these horrific maladies in genetically susceptible individuals. All of the groups of pollutants found in ppm-levels in most or all sewage sludges in EPA's 2009 survey are in the top ten groups of environmental pollutants linked to autism.[16]

The Poop on EPA's 503 Sludge Rule

The 503 sludge rule, which EPA promulgated in 1993, regulates land application of treated sewage sludges. Of all the vast numbers of municipal and industrial pollutants present in sewage sludges, it regulates only nine heavy metals and two nutrients—nitrogen and phosphorus. It does not regulate pharmaceuticals, pesticides, growth hormones, flame retardants, and other potentially harmful pollutants, which EPA strictly regulates in air and water. Many of these chemicals are carcinogenic, mutagenic, neurotoxic, or otherwise harmful to public health and the environment in trace amounts.

In May 1992, the 503 rule failed to pass an internal peer review in EPA's Office of Research & Development (ORD). ORD assistant administrator Erich Bretthauer, however, agreed to let it pass after EPA's Office of Water (OW) promised to fund ORD $2 million per year for five years to address six major gaps in science identified by peer reviewers. The OW promised to revise the rule accordingly. John Walker and others at the OW, however, had no intention of funding opponents to change the rule.

Five months before President Clinton took office, Walker and one of his superiors, Michael Cook, established a cooperative agreement with the Water Environment Federation (WEF). Cook directed a component of the Office of Water called the Office of Wastewater Management.[17] This agreement provided a mechanism for millions of dollars in congressional earmarks to flow from EPA to the wastewater industry's biggest trade association to support a National Biosolids Public Acceptance Campaign.

Upon entering office in 1993, President Bill Clinton appointed Robert (Bob) Perciasepe as assistant administrator for the OW, and the 503 rule was promulgated in February of that year. After leaving EPA in 1998, Perciasepe ran the National Audubon Society's Washington, DC, office, and later served as its COO. President Barack Obama brought him back as deputy EPA administrator in 2009, and appointed him acting administrator at the beginning of his second term in 2013, until Gina McCarthy's confirmation that July.

Under Perciasepe, the EPA-WEF cooperative agreement was amended in 1996 to include a Biosolids Cooperative Research & Development Coordination Project.[18] Its objective was to create a national database of all biosolids-related research projects. Sources of the information included (1) WEF member association leaders and/or biosolids committee chairs; (2) biosolids regulatory agencies; (3) deans of schools of agriculture, engineering, and public health for land grant universities in each state; (4) federal agencies (e.g., USDA, USCOE, USEPA); and (5) other organizations (e.g., AMSA, and the US Composting Council). *WEF deputy executive director Albert C. Grey stated:*[19]

> *A letter will be sent to the appropriate contact requesting the name, phone, address and affiliation of the principal investigators. Each letter will also request a short synopsis of the research program if possible. If there is no response to a letter after six weeks, one or more telephone calls will be made to the contact to follow up.*

Other data to be included consisted of information on the principal investigators' coworkers and affiliations. Grey explained that the final product would be a "compendium of recently completed and ongoing research germane to biosolids."[20] Twenty-five copies would be internally distributed, and the compendium would be promoted by the WEF and sold to others.

According to the amended cooperative agreement, this database was needed to support several milestones of the National Biosolids Public Acceptance Campaign. One was the "development of facts sheets to document and critically examine negative reports regarding the health or environmental impact of biosolids practices in the United States."[21] Apparently, the compendium was used to identify scientists who publish negative reports about biosolids, give EPA and the WEF the opportunity to "out" them and their institutions, and discredit their reports.

In other words, the compendium was a taxpayer-funded, national blacklist of scientists who publish negative reports concerning biosolids and raise concerns about EPA's 503 sludge rule. I imagine people in the biosolids business would pay good money for a list of every scientist doing research on biosolids, who their coworkers are, which institutions they are affiliated with, and whether their work is favorable or unfavorable to government and industry.

In the Decision Memorandum submitted to Michael Cook, the EPA-WEF Cooperative Agreement was described as an "important project" needed to gain "acceptance of the science and the substance of the Part 503 Rule" and overcome "misinformation" spread by opponents.[22] The WEF stated that its purpose was to "provide scientifically credible results that can serve as the basis for future rulemaking efforts by EPA and state agencies."[23] Walker's branch chief, Robert E. Lee, promised that the project would make "beneficial use of biosolids non-controversial by the Year 2000."[24]

Why this agreement was ever approved by EPA's Grants Office is a mystery to me. In my opinion, it clearly violated the Federal Grants and Cooperative Agreement (FGCA) Act of 1977, which prohibits the use of federal grants and cooperative agreements to directly benefit the government. Although violating the FGCA Act is punishable by fines and imprisonment, it has been no deterrence to EPA, USDA, and other federal agencies using massive amounts of taxpayer money to fund grants and cooperative agreements designed to support government policies and silence opposition within the scientific community.

Although it was not apparent at the time, EPA and the WEF planned to achieve public acceptance of the "science and substance" of EPA's 503 sludge rule by almost any means necessary. Their overarching goal was to overcome any "barriers" by the Year 2000.[25] In retrospect, it's clear now that overcoming barriers meant four things: (1) covering up pre-1993 studies that documented adverse effects; (2) discrediting post-1993 "biosolids horror stories"; (3) defunding scientists who oppose the 503 rule; and, finally, (4) creating a body of supportive "science," which EPA scientists call "sludge magic." The first two effects are dealt with below, and the last two are covered in the next two chapters.

Covering Up the Past

Prior to 1993, when EPA passed the 503 sludge rule, numerous studies documented serious adverse health effects and environmental problems from land application of treated sewage sludge. Covering them up wasn't difficult because EPA and USDA had already buried them in internal government reports rather than publishing them in the peer-reviewed scientific literature. For example, EPA's Office of Research & Development (ORD) funded a comprehensive, long-term field study at the University of Florida in the late 1970s to early 1980s.[26] It clearly demonstrated that sewage sludge containing heavy metals at levels allowed under EPA's current 503 sludge rule cause liver and kidney damage in farm animals.

The University of Florida project was a five-year study testing the effects of heavy metals and pathogens in treated sewage sludges from Florida and Illinois on cattle, swine, and poultry. The authors concluded that "certain metals, including cadmium, lead, nickel, and chromium, [are] accumulative in animals consuming forage or grain from sludge-amended soils and therefore have potential hazard to animal health and mankind." Arsenic and molybdenum, which are also now regulated by the 503 rule, were not evaluated in this study. Mean cadmium, chromium, and nickel concentrations in some of the sewage sludges were well below limits of the current (503) sludge rule. They also found that cattle grazing on fields treated with the sewage sludges acquired parasites commonly found in sewage. This study was never published in the peer-reviewed scientific literature. Based on EPA's handling of another government-funded project described below, called the Oak Ridge study, I assume that EPA never released the Florida study for publication because the results challenged the agency's policies on biosolids.

In the late 1990s, scientists at the Oak Ridge National Laboratories (ORNL) conducted similar long-term field studies funded by EPA-ORD.[27] They concluded that even single applications of biosolids could cause long-term ecological damage to forest ecosystems. Despite FOIA requests submitted by environmental activists, EPA kept the study out of the public domain until 1998, when NH Senator Judd Gregg obtained a copy of the Final Report in response to a letter from Caroline Snyder of Citizens for Sludge-Free Land. In 2002, EPA's Office of Inspector General was told that the report existed only in draft form, and was not endorsed by EPA.

In short, the report's authors assessed the effects of treated sewage sludges applied to soils in four major forest ecosystems across the United States. They concluded:[28]

> There is a substantial uncertainty associated with estimates of the quantity of elements that remain in surface soils after a number of years (or for different periods of time in the case of multiple applications).... The bioavailability of elements that were applied in sewage sludge to soils decades ago is not easily estimated. An ecological risk assessment of cumulative loading

*limits for the application of municipal sewage sludge in forests
and rangeland would not be very definitive at this time.*

*A risk assessor could attempt to estimate protective cumulative
loading limits based on multiple lines of evidence (single toxicity,
ambient media toxicity, and field surveys), but such estimates
would also not be definitive. These lines of evidence come from
different ecosystems, soils, sludges, application rates, and organisms, and any estimate of protective loading limits would not be
very precise.*

In other words, ORNL concluded that the uncertainties associated
with long-term bioavailability of heavy metals in sewage sludges are so
numerous that it's not possible to precisely determine what levels are safe
for land application. In laymen's language, the safety of land application of
treated sewage sludge is a shot in the dark. So, it's not surprising that EPA
wouldn't release this report for publication.

What is surprising, even shocking, is that EPA officials apparently lied
to the Office of Inspector General when they said that the report was never
peer reviewed, and existed only in draft form. According to the actual
report (page ii), ORNL's Environmental Sciences Division transmitted the
Final Report to G. Tracy Mehan, EPA assistant administrator for water,
Sylvia K. Lowrance, acting EPA assistant administrator for enforcement
and compliance, and Henry Longest, acting assistant administrator for
ORD, on September 30, 1998. Moreover, a peer review team consisting of
fourteen national experts reviewed the study in 1995 and 1997.

Biosolids Horror Stories

Another major barrier to public acceptance was the growing number of
adverse health effects being reported as land application increased. The
EPA-WEF cooperative agreement refers to such cases as "biosolids horror
stories." An amendment approved in 1994 states:[29]

*This amendment to the Cooperative Agreement will involve a
critical examination of 10 or more unsubstantiated horror stories which have been attributed to the use of biosolids and the*

development of an inventory of beneficial use practices and pro-jects from across the United States.

Unsubstantiated claims of horror stories that have been attrib-uted to the use of biosolids are an important weapon of groups that are opposed to the use of biosolids. WEF will assemble and evaluate information that fully explains what really occurred and translate this information into facts sheets that are readily understandable to the general public.

An internal EPA memo obtained by EPA's Office of Inspector General indicated that Bob Brobst, the EPA coauthor of the Gaskin study described below, did investigate so-called horror stories.[30] The memo states:

Biosolids Horror Stories. *We asked Bob for real life examples of adverse environmental effects from biosolids. Bob sent us a list of sites with groundwater contamination.*

As with the Oak Ridge Study, EPA officials refused to provide the Inspector General's office a copy of the final report on the basis that it was under internal review. Tables of field data attached to the memo, however, indicated that groundwater contamination with nitrates and heavy metals occurred at multiple sites in the following nine states: California, Colorado, Georgia, Illinois, Maine, Minnesota, New Mexico, Nebraska, and South Carolina.

The Gaskin Study

In 1998, I began investigating a growing number of anecdotal reports of illnesses and deaths linked to biosolids. Around 1995, Henry Longest left the Office of Water to become deputy assistant administrator for ORD. Longest, who developed EPA's policies on biosolids in the Office of Water, was now in charge of managing EPA's research scientists, who were critical of those policies.

As part of a settlement agreement in one of my Labor Department cases, Longest transferred me to UGA in 1998 to await termination. Coinciding with this transfer, Bob Perciasepe created a Biosolids Incident Response Team (BIRT) in the Office of Water to investigate a growing number of reports of adverse effects linked to land application of biosolids, including cattle deaths on two dairy farms near Augusta, Georgia. The dairy farms, which had been owned by the McElmurray and Boyce families since the 1940s, were two of the most productive dairies in Georgia. Andy McElmurray explained what happened:[30]

> *We allowed the City of Augusta to apply their sewage sludge to our farmland for 11 years. After several years we began to notice problems, but we could not pinpoint the cause. Two years later, the problems became more serious: a precipitous drop in milk production, sick cattle and excessive cattle mortalities. In 1998, we hired experts to help figure out what was going on with our herd. The experts concluded that the forage which was grown on our sludged fields was contaminated by toxic metals and other industrial pollutants, which were poisoning our cattle.*
>
> *[It's] impossible to manage sludge safely. Over the long term, cattle that graze or consume forage grown on sludged land will suffer from different illnesses because of undisclosed hazardous waste, including heavy metals and mineral imbalances in the forages. The animals will most likely suffer from immune deficiency syndrome, which will not allow them to fight normal cattle diseases. All of this could be very expensive and more than offset any perceived financial gain from sludge application.*

BIRT was comprised of Robert Brobst, who is an EPA scientist in Boulder, Colorado, and two Office of Water employees at EPA headquarters, John Walker and Robert Bastian. According to Perciasepe, BIRT's mission was to investigate "alleged problems associated with biosolids...to provide additional assurances to the public about the integrity and soundness of biosolids management in the United States."[32]

To reassure the public that Augusta's biosolids were safe, Brobst arranged a cooperative agreement with UGA to investigate problems with Augusta's biosolids on the two dairy farms. The Office of Water provided Julia Gaskin, a land application specialist at UGA, with $12,274 to fund the project. Robert Brobst, who headed BIRT, gave her a summary of data (later shown to be fabricated by the City of Augusta) to include when Brobst and Gaskin published their results in a scientific journal.[33] The data provided by Brobst indicated that levels of heavy metals and nitrogen in Augusta's biosolids dropped substantially after EPA passed the 503 sludge rule in 1993.

Together, Gaskin and Brobst published their study in the *Journal of Environmental Quality* in 2003, concluding that Augusta's biosolids "should not pose a risk to animal health."[34] In a national press release issued by UGA, the study's lead author, Julia Gaskin, stated, "Some individuals have questioned whether the 503 regulations are protective of the public and the environment. This study puts some of those fears to rest."[35]

In 2002, BIRT member Robert Bastian provided a draft copy of Gaskin's study to the National Academy of Sciences National Research Council (NRC), which used it in a report to conclude that there is no documented evidence that the 503 regulation has failed to protect public health and the environment.[36] The NRC panel was convened in response to congressional hearings into EPA's retaliations against me and other scientists for documenting problems with the regulation.[37] In 2008, Judge Anthony Alaimo of the US District Court, Southern District of Georgia, found that a plant manager at Augusta's Messerly Wastewater Treatment Plant, Allen Saxon, had fabricated the environmental monitoring reports, which Robert Brobst summarized in Table 2 of the Gaskin study.[38] Judge Alaimo wrote (p. 17):

JUDGE ANTHONY ALAIMO (2008):

> *There is also evidence that the City fabricated data from its computer records in an attempt to distort its past sewage sludge applications.... In January 1999, the City rehired Saxon to create a record of sludge applications that did not exist previously.*

This fabrication occurred just days before Brobst and fellow BIRT member Robert Bastian visited the plant in January 1999. The nexus between Allen Saxon's fabrication of environmental monitoring reports and the arrival of Brobst and Bastian to collect these records suggests that the data, which Bastian provided to the NRC panel, may have been fabricated specifically for publication in the Gaskin study, to support the efficacy of EPA's 503 sludge rule and to dismiss any link between biosolids and the cattle deaths on the two dairy farms.

Falsifying data required under the Clean Water Act is a criminal violation punishable by fines and imprisonment. However, the Justice Department showed no interest in prosecuting anyone over the fabricated data. Attorney Ed Hallman also represented two of the dairy farmers and me in a qui tam case.[39] In my sworn testimony in this case, I recalled that in 2003 I testified as an expert witness at public hearings before the City and County Council of Honolulu, Hawaii. The council was considering whether to move forward with a contract submitted by Synagro Technologies, Inc., the same company that spread Augusta's biosolids in 1999 during the Gaskin study.

I recommended that the council have the University of Hawaii perform some simple tests to determine the presence and potential for regrowth of pathogens in samples of Synagro's product before approving the contract. The purpose was to validate the company's claims that its product was sterile and presented no risk of infection. These tests would require delaying the contract approval for several weeks. Alex Strauss, a division director for EPA Region 9 in San Francisco, responded by threatening Honolulu with over $5 million in fines for each day it delayed Synagro's contract.[40]

Muzzling the Messenger

As indicated above, Henry Longest, EPA's acting assistant administrator for ORD, and other EPA officials accused me of violating the Hatch Act and government ethics rules by publishing my 1996 *Nature* commentary, "EPA Science: Casualty of Election Politics," and criticizing EPA's sludge rule in editorials published by a local newspaper.[41] Longest also cut off my internal EPA funding, and offered to transfer me to UGA for four years if I would agree to retire afterward. I agreed because it was my only option for continuing my research on biosolids.

When I published a research article in *Nature* in 1999 that raised new concerns about EPA's sludge rule, EPA assistant administrator Norine Noonan ordered the removal of my EPA laboratory director for approving the article. Then, in 2001, the director of EPA's Office of Wastewater Management, Michael Cook, and one of his employees, John Walker, met with two Synagro executives, Alvin Thomas and Robert O'Dette, over lunch to discuss another research article I wrote, which was undergoing an internal review at EPA.[42] It linked Synagro's biosolids to illnesses and deaths at multiple sites across the United States. Walker also met separately with O'Dette over breakfast, and asked Synagro to provide information he could use against me in his internal EPA peer review of my research article.

Synagro followed up several months later by emailing Cook, Walker, and other EPA officials a "white paper" containing allegations of research misconduct against me and one of my coauthors at UGA. Specifically, Synagro alleged that EPA had never approved my research papers on biosolids, and that my work on biosolids was not peer-reviewed. Walker knew the allegations were false. In his peer review, he stated:

I have attached my review of this manuscript along with a completed peer-review form. Unfortunately this paper has received an OK for publication by Mr. Lewis' supervisor.... As Alvin Thomas explained at the WEF meeting in Anaheim, California, one year earlier, this meant that I was misusing federal and state funds—potentially a federal crime punishable by fines and imprisonment. Walker immediately forwarded Synagro's white paper to a waste disposal company in Atlanta, Georgia, for public distribution under EPA headquarters letterhead.[43] Synagro filed its allegations with UGA as a formal petition to investigate research misconduct, and hired Georgia senator Kasim Reed, now mayor of Atlanta, to pressure UGA not to dismiss the petition.[44]

It was discovered later that Walker's peer-review comments were written by a USDA scientist associated with Rufus Chaney's efforts to promote biosolids.[45] Walker copied the lengthy review verbatim and passed it off as his own. Fortunately, the other peer reviewers recommended publication. So, the paper passed two internal peer reviews, first in the Office of Research, and then in the Office of Water. Published in *BMC Public Health*, it was the first paper to document illnesses and deaths linked to biosolids in the peer-reviewed scientific literature.[46]

Synagro and NEBRA published Synagro's white paper on their websites; then Synagro and the WEF began writing letters to EPA administrator Christie Whitman and others requesting that EPA investigate me for research misconduct and potential criminal misuse of federal and state funds.[47] In 2003, Synagro filed its white paper allegations with UGA as a formal petition to investigate research misconduct, and UGA forwarded it to EPA.

In 2013, the Justice Department collected $35,000 to settle over $61,000 in court costs, which the dairy farmers and I owed to defendants in our qui tam lawsuit.[48] The assistant US attorney in Macon, Georgia, informed my attorney, Ed Hallman, that EPA specifically asked him to collect the court costs from me, saying, "We think David Lewis has money." This probably explains why the Justice Department demanded that I, and only I, estimate the current value of all of my assets. EPA, in other words, used the Justice

Department to send a powerful message to scientists who may be thinking about filing qui tam lawsuits over research fraud that any of its employees may commit in the line of duty.

The $35,000 represented only a small portion of over $250,000 UGA paid private law firms to defend against a qui tam lawsuit that the dairy farmers and I filed.[49] The purpose of our lawsuit was to compel UGA to withdraw the fabricated data that EPA and UGA published for a study that EPA and the National Academy of Sciences used to dismiss biosolids as having caused cattle deaths on the two dairy farms near Augusta. In July 2013, we sought information on the amount EPA paid private attorneys to defeat our lawsuit and avoid having to withdraw the fabricated data.[50] Nine months later, we're still waiting on EPA to process our FOIA request.

SLUDGE MAGIC

Dr. Alan Rubin, a career chemist at EPA's Office of Water (OW), is considered the primary author of EPA's 503 sludge rule, which allows treated sewage sludge, aka biosolids, to be land-applied to farms, forests, parks, and other private and public lands.

Rubin was also one of the scientists at EPA headquarters in Washington, DC, involved in retaliations against me and others who reported adverse health effects associated with biosolids in the scientific literature or the news media. *Time* magazine (September 27, 1999) ran a short article about Rubin attacking me and mailing "death threats" on EPA letterhead to private citizens concerned about biosolids, saying, "Ask not for whom the bell tolls; it tolls for thee!" When deposed in my Labor Department cases, Rubin explained what motivated his attacks:[1]

RUBIN QUESTIONED BY MR. KOHN (1999)

Q. Are you proud of the work you did?... Do you feel, in any way, hurt or upset to have someone like Dr. Lewis criticizing it?...Professionally hurt, a little?

A. Somewhat.

Q. How so?...

A. Well, I think my professional reputation, to a large extent, is based on my association with biosolids, 503 and its technical basis. So I feel my reputation would be somewhat disparaged if the basis of the rule, and the scientific findings were shown to be in error.

The term "sludge magic" originated when EPA's 503 sludge rule was undergoing internal peer review at ORD in 1992. Dr. Robert Swank, the research director at the EPA lab in Athens, Georgia, where I worked, called Dr. Rubin, one of Walker's coworkers at EPA headquarters in Washington, DC. At that time, there were no studies published in the scientific literature that supported the theory that organic chemicals and other industrial pollutants are sequestered by sewage sludge and cannot be taken up by plants and animals. At least Rubin couldn't produce any. When Swank asked him to explain how it worked, Rubin replied, "It's magic."

When asked where all the studies supporting "sludge magic" could be found, Rubin deferred to USDA agronomist Rufus Chaney:[94]

RUBIN QUESTIONED BY MR. KOHN (1999)

Q. You called it sludge magic?

A. Yes, that is my term. "sludge magic" [means] there are unique properties in the biosolids matrix that sequester metals, that sequester organics. By sequester I mean significantly reduce the mobility to move from the biosolids out to the environment, and the matrix is really complex, and has organic material in it, organic pollutants, I'm talking about organic materials, like unit type materials, and carbohydrates, and manganese, and iron, and phosphorus, and all of these work together with the soil in a matrix to significantly reduce, if not eliminate movement of pollutants from the biosolids out to the environment. The processes, some of them are understood, some of them are not that well understood, but the whole thing taken together is called magic. So I coined the term magic.

Q. And the "sludge magic" which prevents harmful stuff that is in the sludge escaping the sludge?

A. Moving at any significant flux or rate out to the environment to create doses of pollutants that would harm plants, animals or humans.

Q. ...these studies [are] kept somewhere?

A. No, they are actually—well, Chaney is probably the one that has them all, he is like a walking encyclopedia.

Usually, sewage sludges are converted into biosolids by simply adding lime to raise the pH, which reduces odors and pathogen levels. Lime doesn't destroy most pollutants found in biosolids. Some pollutants, such as the pesticide malathion, are even converted to much more highly toxic pollutants at high pH. After working in EPA's biosolids program for over thirty years, Dr. Rubin still couldn't explain how biosolids prevent pollutants from posing a serious risk to public health and the environment. To be honest, I never cared for magic. No matter what a magician says or does, it's all based on deception. So, I looked forward to my attorney, Ed Hallman, deposing Dr. Chaney at USDA's Animal Manure and By-Products Laboratory in Beltsville, Maryland.

Rufus Chaney

Dr. Chaney made it clear that we had come to the right person. The people at EPA, he testified, have never understood the science of biosolids, and he has to reeducate them time after time.[3]

CHANEY ANSWERING MR. HALLMAN (2009)

EPA withdrew the original proposed rule and completely rewrote it. Actually I played a very significant role in what the rule became. It's evident in the record. And even at the end I provided comments through USDA, approved at higher levels, saying that the rule needed a few more revisions before it was issued. But, yes, I was heavily involved in bringing to fore the science about biosolids that needed to be the basis for the rule.

Chaney explained that unique properties prevent pollutants from becoming *bioavailable*. That means that they aren't taken up or absorbed by plants and animals, and pose little or no risk to public health or the environment no matter which pollutants are present, or what their concentrations are. This holds true, he testified, regardless of whether wastewater treatment plants are even working properly. In other words, it *really* is magic.

Rufus Chaney (C), Scott Angle (R). USDA Photo 1995.

One of Rufus Chaney's primary collaborators, Jay Scott Angle, replaced Gale Buchanan as the agricultural dean at UGA in 2005, the year we filed our qui tam lawsuit over the Gaskin study.[4] President Bush appointed Buchanan under secretary of agriculture for research, education and economics the following year.[5] Before leaving for Washington, Buchanan called one of my colleagues and said the stink over the Gaskin study was causing a big problem. Two years earlier, the director of UGA's School of Marine Programs testified that he was advised not to hire me as a faculty member "because we're dependent on this money...grant and contract money...money either from possible future EPA grants or [from] connections there might be between the waste-disposal community [and] members of faculty at the university."[6] In a press release announcing Angle's appointment, UGA president Michael Adams and provost Arnett Mace applauded Angle's research on biosolids at the University of Maryland, which discredited earlier reports that treated sewage sludge negatively affected nitrogen fixation in soil.[7]

The part about wastewater treatment plants not working properly, however, isn't magic. Many wastewater treatment plants throughout the United States aren't working properly, and are constantly in need of being repaired or upgraded to keep up with population growth. To help with this problem, EPA created a revolving loan program under the Clean Water Act to pump billions of dollars into the states to keep their wastewater treatments plants pumping properly. Chaney reasons that because the system as a whole is in constant need of repair, and there are still no documented cases of adverse health effects in the peer-reviewed scientific literature, "sludge magic," as Rubin calls it, works even when wastewater treatment plants don't.

Chaney further reasoned that any peer-reviewed scientific articles claiming that land application of biosolids poses a risk to public health or the environment must be false because no scientists funded by the US government and other reputable institutions have documented adverse effects from biosolids since the 503 sludge rule was passed in 1993. Chaney, I'm sure, would dismiss any notion that a lack of documented cases since 1993 has anything to do with EPA's establishing a National Biosolids Acceptance Campaign to fund land grant universities, such as UGA, to promote biosolids and eliminate scientists who report problems or document adverse effects.

In 1992, EPA's sludge rule failed to pass a scientific peer review within its Office of Research & Development (ORD) where I worked. Chaney blamed scientists in EPA's Office of Water (OW) for this failure:[8]

CHANEY ANSWERING MR. HALLMAN (2009)

They originally proposed a rule where they even had the data screwed up. I don't know how much you know about that. But the original rule would have essentially prohibited all land application.... They had an uptake of PC and B instead of PCB, taken up a thousand times easier than PCBs, and so the limit would have been a thousand times lower than it needed to be. So there were lots of errors the first time around, stupid errors. They didn't—they didn't review it with USDA or Food and Drug Administration before they put it on the street and they suffered and had to withdraw it and start over.

In his deposition, Chaney pointed out that adverse health effects from biosolids were documented in the scientific literature before 1992, and that he himself authored many of those studies.[9]

CHANEY QUESTIONED BY MR. HALLMAN (2009)

Q. And you believe that all the studies you've seen, including the ones that you have coauthored and worked on, indicate that the land application of sewage sludge in accordance with 503 is safe...and is not a danger to human health and welfare, is that correct, if it's applied in accordance with those regulations?

A. I won't disagree with that. I had advised EPA that I wanted a lower cadmium limit.... I won the battle because pretreatment and the universal understanding of the unacceptability of cadmium in biosolids has led to biosolids declining to 1 to 2 ppm in most cities in the United States. Biosolids has become remarkably less contaminated because of what we've done with the 503 and because of the publications, such as mine, which showed adverse effects of previous practices.

The phenomenon by which biosolids have become less contaminated with cadmium is clearly evident in the data that the City of Augusta reported to EPA and the State of Georgia (see Appendix II: Biosolids Cadmium Data Pre- and Post-1993). These are the same data that EPA and UGA published in a study used by the National Academy of Sciences to conclude that Augusta's biosolids were not responsible for hundreds of cattle that died on two dairy farms where it was applied. The data purportedly show that monthly cadmium levels in the city's sewage sludge fluctuated wildly up to 1,200 ppm from January 1980 to February 1993, the very month that EPA promulgated the 503 rule. Then, one day in the middle of February, cadmium levels suddenly dropped to 5 ppm and remained at that level from then on.

Dr. Chaney wants everyone to believe that high cadmium levels in biosolids made people and animals sick all across the country before February 1993 when EPA passed the sludge rule under his guidance. Then, after EPA followed his instructions, no human or animal has since gotten sick from cadmium in biosolids. All I can say is that it takes some really powerful magic to make every industry in America, which has been disposing of large quantities of cadmium since the Industrial Revolution, drop it to a few ppm on the same day in 1993 and hold it at this level. It's even more amazing because no regulatory agency at the state or federal level monitors levels of cadmium, or anything else, in biosolids.[10] They just take whatever data the cities provide.

In Augusta's case, we happen to know that the city's "sludge magic" was fake because the city's former plant manager, Allen Saxon, confessed when deposed by Mr. Hallman. Judge Anthony Alaimo explained exactly how it happened when he ruled in favor of the McElmurray family in a lawsuit, and ordered Dr. Chaney's agency, the USDA, to pay for crops the family couldn't plant because their land was too contaminated with cadmium and other hazardous wastes in Augusta's biosolids. Judge Alaimo wrote, "In January 1999, the City rehired Saxon to create a record of sludge applications that did not exist previously."[11]

The year 1999, by the way, is the same year EPA gave UGA a federal grant to publish Augusta's data in the Gaskin study. Then, as soon as Mr. Saxon finished making "sludge magic" happen, all of the original data Augusta reported to the Georgia Environmental Protection Division (EPD) between 1993 and 1999 magically disappeared, and not just in Augusta. They turned up missing from the EPD records in Atlanta as well. EPA doesn't know what happened to the data, nor does the EPD, nor the City of Augusta, nor UGA. All of the data just magically disappeared from city and state records at the same time cadmium disappeared from Augusta's sewage sludge. Apparently, no one other than us had any interest in learning how Augusta's records of high cadmium concentrations miraculously disappeared from Augusta and the Georgia EPD in Atlanta in January 1999, or how high cadmium levels magically disappeared at wastewater treatment plants across America in February 1993.

According to Chaney, it just doesn't matter anyway whether the data are fake or real. He explained in his deposition:[12]

CHANEY QUESTIONED BY MR. HALLMAN (2009)

Q. Ms. Gaskin could have totally made up all that data and you would still rely on it because it was in a peer-reviewed study; is that accurate?

> *A As long as it—as long as it was in general agreement with*
> *general patterns established in hundreds of papers....*

To sum up Chaney's position, because Gaskin's paper concluded that Augusta's sludge did not pose a health risk, it's valid research even if all of the data are fabricated. On the other hand, people should disregard scientists who report problems with biosolids, even if their work is published in the peer-reviewed scientific literature. That's because researchers at universities funded by government and industry to support biosolids assure us that the practice is safe and have published hundreds more papers saying that there are no problems.

BMC Public Health

In 2004, Chaney commented on the US Composting Council's (USCC's) list serve about my termination by EPA acting assistant administrator Henry Longest, who developed EPA's sludge policies in the late 1970s.[13] USCC is currently headed by Lorrie Loder, Synagro's product marketing director. Chaney, of course, supported Longest's decision to end my career in environmental research for publishing research that raised public concerns over biosolids. He contrasted my *BMC* study with the Gaskin study:

CHANEY USCC (2004)

> *The paper by Gaskin et al. [Gaskin, J. W., R. B. Brobst, W. P. Miller, and E. W. Tollner. 2003. Long-term biosolids application effects on metal concentrations in soil and bermudagrass forage J. Env. Qual. 32:146-152.] reports objective measurements on the soil metal concentrations, and metals in forages growing on the soils....*

> *[Lewis's] publication [Lewis D. L., D. K. Gattie, M. E. Novak, S. Sanchez, and C. Pumphrey. 2002. Interactions of pathogens and irritant chemicals in land-applied sewage sludges (biosolids). BMC Public Health 2:11.] contains none of the data from examination of biosolids exposed subjects, and lacks the*

comparison with randomly selected individuals from the general populations. It is not valid epidemiological science....

I support the whistle-blower rule and process as strongly as any other citizen or government employee. I happen to believe that Dr. Lewis has been treated fairly. Claims and opinions about public health are not peer-reviewed scientific evidence. EPA and other agencies have to base rules on the peer-reviewed papers, and to consider the weight of evidence. Some papers are more complete in proof of the issue tested, as I noted above regarding proof that some source caused a specific human infection.

Chaney's remarks serve to underscore an important observation worth mentioning. Scientific journals overlook government scientists' job-related conflicts of interest.

Rufus Chaney is paid to promote government policies on biosolids. If he didn't support them, he would lose his job, just as I did, only much more quickly. Chaney is the government's number-one scientist charged with creating and defending whatever "science" is needed to support the sludge regulation, which he developed for EPA to enforce. Consider the fact that *every* municipality and industrial polluter depends on Rufus Chaney to fill the scientific literature with studies "proving" that dumping their hazardous wastes in sewers is environmentally beneficial and doesn't harm public health. His position at USDA speaks for itself:[14]

CHANEY ANSWERING MR. HALLMAN (2009)

I've been appointed in a category which is above GS-18 called senior scientific research service. Within that, there are no sub-grades. There is a group —there is only about ten of us in all of my agency that have reached that level.... I would say I'm the US Department of Agriculture's most knowledgeable scientist about biosolids.

How much trust should be placed in a scientific article published by a scientist working for GlaxoSmithKline (GSK) concluding that a GSK product is safe? Such an article would have little, if any, credibility. On the other hand, how much trust would the article merit if it concluded that GSK's product is unsafe? Quite a lot, I would say, because the scientist would be sacrificing his or her career to publish it.

Before Chaney developed EPA's sludge rule, his papers concluded that cadmium in sewage sludges applied to land presented a risk to public health. But after the sludge rule passed, Chaney's position changed. Now he argues that biosolids are safe, no matter what. For example, it doesn't matter whether Augusta, Georgia, covered up high levels of cadmium in its sewage sludges, or even if its wastewater treatment plant is functional. So far as Chaney's innumerable publications are concerned, I think that the only ones that are trustworthy are those he published before he got into the business of defending the sewage sludge regulations he developed for EPA.

Unfortunately, publishers of scientific journals, for whatever reason, don't pay attention to job-related conflicts of interest held by government scientists. As a result, we have a body of science behind EPA's sludge rule, which EPA can explain only as magic, and it doesn't even matter whether data are fabricated. The standard for quality control and research ethics cannot possibly go any lower.

When the dairy farmers and I filed a qui tam lawsuit over EPA and UGA employees using a federal grant to publish fake data, EPA and UGA hired private attorneys to protect the fabricated data, which the National Academy of Sciences used to support EPA's regulation. This level of institutional corruption threatens scientific integrity throughout government and industry. We have no mechanisms in place to even confront it, much less stop it. The only honest thing left to do is throw out the whole body of "biosolids science," and start over with a new group of scientists who have no conflicts of interest with government and industry.

Combating institutional research misconduct requires carefully reading the original source documents.

Take, for example, Chaney's statement that our *BMC* article contains none of the data from examination of biosolids exposed subjects, and lacks the comparison with randomly selected individuals from the general populations. It is not valid epidemiological science. . . . These claims originate with Synagro's white paper containing allegations of research misconduct against me and my coauthors. These allegations represented a failed attempt to have UGA withdraw our study published in *BMC Public Health*, which linked Synagro's products to illnesses and deaths. Our study was the first to document adverse health effects from biosolids in the peer-reviewed scientific literature. In 2004, Synagro withdrew its allegations after UGA forwarded its petition to EPA, and EPA dismissed the allegations.[15]

The false allegations that Chaney parroted appear throughout Synagro's white paper, which states, for example:[16]

SYNAGRO WHITE PAPER (2001)

[Steps Lewis should have taken] include analysis of biosolids composition, fate and transport of chemicals and pathogens, determination of dose-response relationships, and methodology for and identification of the cause of health ailments purportedly associated with an environmental contaminant....

Such studies should involve a comparison of outcomes for subjects who are exposed to biosolids (treatment groups) and other subjects who are not exposed (control groups).... The leading study, a comprehensive multi-year study of Ohio farm families living near land-applied fields, reported "no adverse health effects...in either people or animals." (Cit. 38.) While Dr. Lewis admitted that this study was based on sound epidemiology, he refuses to apply its techniques....

Our *BMC* paper, which is available online at www.biomedcentral.com/1471-2458/2/11, *does* contain this information. It includes, for example, data we obtained from the patients' medical records, and a dose-response analysis of exposed and unexposed individuals in an area near a field treated with biosolids (Figure 2).[17] This, in fact, was the field near Shayne Conner's house, where he died from respiratory failure.

Conner's parents sued Synagro, which bought out the company that applied the biosolids. EPA ethics officials approved of my serving as an expert witness for plaintiffs, and required that I donate any expert witness fees to EPA or other governmental or nonprofit organizations. It was the only way I could obtain medical records and other important documents involved in the case. Because the *BMC* article challenged the safety and efficacy of EPA's sludge rule, it did not represent a conflict of interest on my part insofar as being biased toward supporting EPA's interests. My settlement agreement over my 1996 *Nature* commentary left EPA with the option of not terminating me when my appointment to UGA ended. EPA, therefore, made sure that I remained financially motivated to support its policies on biosolids while working at UGA.

By serving as an expert witness, I was able to obtain data from all but one household in Conner's area. Random selection of test subjects, which Chaney criticized us for not using, is employed when researchers can only test a small portion of a population. It ensures that a small sample group is representative of the whole population. We reported:

LEWIS ET AL. BMC PUBLIC HEALTH *(2002)*[18]

Based on a least-squares analysis, proportions of individuals with symptoms increased linearly from 40 to 80 h (r^2 0.98) with time exposed to wind blowing from the field; all occupants in households with exposure $\geq$ 80 h reported symptoms (Fig. 2). Proportions of individuals with symptoms also decreased linearly with distance from the field from 130 to 320 m (r^2 0.95); all occupants in households living $\leq$ 130 m from the field reported symptoms.

To obtain data on the residents' symptoms, we used questionnaires patterned after the Ohio study recommended in Synagro's white paper. As *Nature* reported in an editorial and news article, a multi-university follow-up study in Ohio published in 2007 independently confirmed our findings, and cited our work published in *BMC Public Health*.[19]

To give Dr. Chaney the benefit of the doubt, I'll assume he has never actually read our paper. Otherwise, I would have to say he just outright lied about what was, or was not, included in the way of data. Gaskin, her two UGA coauthors, and the original UGA principal investigator on her study voiced the same accusations after Synagro filed its white paper allegations as a research misconduct petition and sent Gaskin a copy. To address the allegations, I arranged for Gaskin's group and her department head to meet with my group and my department head in the Department of Marine Sciences.

Before discussing our paper, I asked for everyone who had read it to raise their hands. Everyone associated with Gaskin's project had coauthored an editorial criticizing our paper, which Gaskin had prepared to submit to an industry publication. Gaskin, however, was the only person in her group who claimed to have actually read our paper. The reason I asked for a show of hands is because I knew that Gaskin's editorial, like Chaney's comments published by USCC, were drawn from the industry white paper.

I assumed that Gaskin had, in fact, read our paper. But then I began to wonder when she leaned across the table at me to raise her voice and drive home the one point she seemed convinced was my Achilles' heel. "Why didn't you report that you just looked at Class B biosolids?" she asked. I stared in dismay at both her demeanor and the question, then thumbed through our paper to try to locate the information for her. Under the sub-heading "Assessing Environmental Conditions" in the "Methods" section, our paper states:

LEWIS ET AL. BMC PUBLIC HEALTH (2002)[20]

County records indicated that biosolids-related complaints for individual patients described in this study were concurrent with land application of Class B biosolids. In the case of one family (Residents 19–22), however, records indicated that dairy wastes, rather than biosolids, had been applied.

According to EPA's 503 sludge rule, Class A biosolids have no detectible pathogens, whereas low levels of indicator pathogens are permitted in Class B biosolids. Of course, pathogen levels are pretty meaningless when sewage treatment plants aren't equipped to detect most pathogens that survive waste treatment processes. Moreover, most bacterial populations that are killed back can re-establish themselves within a few days after biosolids are stockpiled, or spread on land.[21]

It's like cooking the Thanksgiving turkey. Eating it fresh out of the oven is one thing, but after it's been sitting out for a few days is a different matter. Biosolids are rich in proteins, which allow staphylococci to proliferate just as they do with turkey dinners. In our study published in *BMC Public Health*, we discovered that one out of four residents who reported irritation of the skin, eyes, or respiratory tract from exposure to biosolids had staphylococcal infections involving *S. aureus* or *S. epidermitis*. Two of the three deaths linked to biosolids were caused by *S. aureus* infections. Because multi-antibiotic-resistant bacteria are common at wastewater treatment plants, biosolids-related infections are of particular concern.[24]

During her depositions, Julia Gaskin was often combative and rude, but always honest. She testified that she believed Augusta's biosolids harmed the McElmurray and Boyce dairy farms, and she said that her study included plenty of data that supported their lawsuits.[23]

GASKIN ANSWERING MR. HALLMAN (2009)

A. Now, you have characterized that the EPA has used this against them. There is certainly data in here that could have been used to support them as well.

Q. What data?

A. The fact that we had high cadmium and molybdenum in three fields that had been—and forages in three fields that had been greater than six years. The fact we saw a reduction in copper and molybdenum ratios with long-term biosolids application.

She admitted that she knew there were problems with Augusta's data, which EPA gave her to publish, and had no reason to argue with Judge Alaimo's ruling that the data were fabricated.[24]

Perhaps most revealing, Gaskin and her EPA coauthor, Robert Brobst, testified that they both objected to Tracy Mehan, EPA's assistant administrator for water, using their study to discredit the McElmurray and Boyce cases. Specifically, Mr. Hallman questioned them about a letter Mehan issued on December 24, 2003, in which Mehan used the Gaskin study to disregard problems on the two dairy farms. In that same letter, Mr. Mehan cited the National Academy of Sciences, National Research Council (NRC) report concluding that there was no documented evidence that biosolids applied under EPA's 503 sludge rule has ever harmed public health. The NRC report cites a draft version of the Gaskin study, which Robert Bastian at EPA headquarters provided to the NRC for the purpose of disregarding the McElmurray and Boyce cases. Attorney Ed Hallman read Brobst's testimony, then questioned Gaskin:[25]

BROBST, GASKIN QUESTIONED BY MR. HALLMAN (2009)

[Brobst]

We, the authors, at least Julia and I, will stand by that the study had nothing to do with the dairy farms. I mean, we both said that on several occasions, and I believe we will both stand by that. And I have conveyed that to headquarters. If they choose to not listen or choose to listen, that's up to them. I don't have any say in how they make these paragraphs and how they form things and form their conclusions. I wouldn't have done it that way.

[Gaskin]

Q. Do you recall any conversations that you've had with Mr. Brobst about the study had nothing to do with the Boyce and McElmurray farms?

A. Yes.

Q. Tell me the substance of those conversations.

A. I, the substance of the conversations were concerns that our study was being used, that people were citing our study as if the dairy farms were part of what we had sampled, and they were not. And I had concerns about that, that even though the JEQ *article clearly said beef cattle farms, that some people were not being clear about that fact.*

Q. Did you ever voice those concerns to anyone besides Mr. Brobst?

A. I voiced those concerns to Mr. Brobst and also at one point Ned Beecher.

Q. Who is that?

A. He is the director of the Northeast Biosolids Association.

Q. What did you tell him?

A. I told him that I was concerned that the JEQ article was being conflated with the dairy and that our study did indicate that there was not a widespread problem, but it did not specifically address the dairy concerns.

In the end, therefore, it appears that EPA headquarters in Washington, DC, set up UGA to publish scientific data that would mislead the National Academy of Sciences to conclude that there was no documented evidence that biosolids applied under EPA's 503 sludge rule has ever harmed public health.

NRC Report (2002)

Responding to the congressional hearings into EPA retaliations against me, my EPA laboratory director, and others who have questioned the science EPA uses to support its sludge rule, EPA called upon the National Academy of Sciences National Research Council (NRC) to reevaluate its scientific basis. Ellen Harrison, an NRC panel member from Cornell University, provided the panel with copies of my unpublished manuscripts and two in-press, peer-reviewed journal articles (*BMC Public Health*, 2002; *ES&T*, 2002). Harrison, director of the Cornell Waste Management Institute, and her coauthors published a well-documented, peer-reviewed article concluding that EPA's current sewage sludge regulation does not protect human health, agriculture, or the environment.[26] She was also part of a group of NRC panel members selected to brief EPA on the academy's findings when their report was electronically released on July 2, 2002. She testified in my labor case:[27]

HARRISON QUESTIONED BY MR. KOHN (2003)

Q. I'm looking for a larger-picture question here, what would you state would be Dr. Lewis' major contribution in terms of the concerns he was raising to the National Academy review process?

A. David is the only scientist that to that time had raised the scientific issues that might lead to exposure and disease and so David's ideas in that regard, I think, were important to sort of

framing the National Academy panel's in recognizing that...
there are a lot of gaps here, there are plausible routes of expo-
sures that we haven't assessed. So David's role was—I mean
in my book David was a hero in this regard basically. Despite
the incredible flack he was getting, [he] put forward reason-
able scientific theories, backed by some research to suggest that
there were plausible routes of exposure and that in fact illness
might be resulting. He, I mean as far as I'm concerned, he kind
of turned the whole thing around...I think without David's
involvement we wouldn't be at all where we are today in terms
of looking at the safety issues anew. David raised-David gave a
legitimacy to the allegations that has made it impossible to ignore
the alleged health issues.... So I think David has probably been
the most important player in all this.

Although the report drew heavily upon my unpublished manuscripts, the electronic version only cited one paper, an *ES&T* article. Susan Martel, an NRC staff member, explained to Harrison that all but one reference to my work were removed from draft versions of the NRC report based on input from panel members. Then, according to Martel, the panel chair, Thomas Burke, removed the one remaining reference to my ES&T article from the final copy of the report, which is posted on the NRC's website.[28] Burke took that action after he and Martel received the following email from panel member Greg Kester, the biosolids coordinator for the State of Wisconsin:[29]

Hi Tom and Susan—In contrast to your message that the brief-
ings went well, I am quite disturbed by what I have heard tran-
spired at the EPA briefing this morning. Among other items, I
heard that EPA staff in the biosolids program were referred to
as "the usual suspects" and basically denigrated for their work in
the program. The message was also taken that their work should
be devalued and the work of David Lewis should be elevated. I
did not agree to such representation nor do I believe much of the
committee did. We specifically noted that EPA should not be crit-
icized for the work they did.... While EPA may not have been
moved by the criticism, there are those on the Hill who would
love nothing more than to criticize EPA.

One year earlier, Synagro VP Robert O'Dette had emailed Kester a copy of his white paper accusing me of research misconduct. Kester, in turn, forwarded it to senior officials at EPA headquarters and other EPA offices throughout the country.[30] In his email, Kester stated:[31]

> *This paper presents many of the issues raised by Dr. Lewis in the New Hampshire case and provides compelling refutation. It was written by Bob O'Dette of Synagro.*

The NRC panel used Synagro's white paper in its deliberations over my research, and rewarded O'Dette by using a photo he submitted for the cover of the NRC report. Although the panel liberally borrowed from my unpublished and in-press papers without citing the source, it was careful to credit O'Dette as the source of its cover photo. Then, after removing my in-press, peer-reviewed articles documenting scores of cases of adverse health effects across the country, the NRC panel falsely reported: "There is no documented scientific evidence that the Part 503 rule has failed to protect public health."[32]

In 2003, my coauthor and I submitted our work to *Environmental Health Perspectives* (*EHP*). We were shocked when the editor rejected it on the basis that we did not credit the NRC for *our* work. The editor wrote:[33]

> *A major shortcoming of the manuscript is the lack of any reference to the very recent National Research Council report, Biosolids Applied to Land, which addresses virtually all of the issues raised by the authors. A Commentary on that report would give the authors ample scope to present their views, either criticism or support, in a contemporary context.*

Fortunately, after I provided extensive documentation proving that we were indeed the original sources of all of the information in our submission, *EHP* accepted the paper. The editors allowed us to reference our own original works, and just credit the NRC with "echoing" our findings and recommendations.[34] But the fallout from what the NRC panel did wasn't over yet.

In 2008, a *Nature* reporter called me wanting my response to a federal judge, Anthony Alaimo, ruling that data in the Gaskin study were fabricated to cover up cattle deaths linked to hazardous wastes in Augusta's sewage sludge. *Nature*, as it turned out, was putting together a two-page news article and editorial about our research at UGA, pointing out that a multi-university study in Ohio had independently confirmed our findings. I thought to myself, *it doesn't get much better than this.* Then I read the editorial:

NATURE *EDITORS (2008)*[35]

In what can only be called an institutional failure spanning more than three decades—and presidential administrations of both parties—there has been no systematic monitoring programme to test what is in the sludge. Nor has there been much analysis of the potential health effects among local residents— even though anecdotal evidence suggests ample cause for concern. In fact, one of the studies used to refute potential dangers, published in the Journal of Environmental Quality *in 2003 by researchers at the University of Georgia in Athens, has been called into question (see page 262).*

Even the National Academy of Sciences seems to have been taken in. A 2002 report from the academy cited the then unpublished Georgia work as evidence that the EPA had investigated and dismissed claims that sewage sludge had killed cattle, but the study had not looked at the dairy farms in question. And although it may be technically true that there was no documented evidence of sludge applications causing human illness or death, the academy also cited work by an EPA whistleblower, David Lewis, suggesting at least an association between these factors.

If anything, recent research underscores those findings. The Georgia citation notwithstanding, the academy did outline a sound plan for moving forward. It recommended among other things that the EPA improve its risk-analysis techniques; survey the sludges for potential contaminants; begin tracking health complaints; and conduct some epidemiological analyses to determine whether these reports merit concern.

To read the NRC report, the *Nature* reporter located it on EPA's website rather than using the NRC website. After I filled the reporter in on what happened, *Nature* ran the following correction, which contained more misinformation, which the NRC provided to *Nature* in an attempt to explain why it removed the last remaining reference to our work in the final version of its report.

NATURE *EDITORS (2008)*[36]

*Correction: The 2002 biosolids study from the National Academy of Sciences (NAS) did not reference research into health impacts by Environmental Protection Agency (EPA) whistleblower David Lewis, as reported in our News story "Raking through sludge exposes a stink" (*Nature 453, 262–263; 2008). *The citation was included in a prepublication draft that is still posted on the EPA's internet site, but the NAS panel voted to remove the reference before final publication. An NAS spokesman said the panel decided the information was not relevant as the panel was not charged with evaluating health impacts.*

For whatever reason, EPA only posted the electronic version of the NRC report containing the one surviving reference to our work. I like to think that, in the end, EPA couldn't stomach the way it all turned out, and decided to give me and my coauthors some credit for what we did. At least panel member Ellen Harrison got in the last word about the National Academy of Sciences removing the last remaining reference to our work:[37]

HARRISON TO NATURE *EDITORS (2008)*

The NAS made this change to the report without permission from the panel. This is a violation of the NAS procedures requiring full committee consensus on reports. I would not have approved the removal of this reference since it was clearly relevant to the work of the committee....the unilateral action of NAS to remove the reference was highly inappropriate.

These People

EPA officials and WEF officials involved in the National Biosolids Public Acceptance Campaign systematically funded scientists who supported the 503 sludge rule while eliminating those who did not. In 2002, a Texas county commissioner invited me to speak at a public hearing about a growing number of illnesses linked to Synagro's biosolids in his area. I agreed on the condition that he invite Synagro to have its own expert rebut my arguments. So, the commissioner wrote a letter to the company's VP for government relations, Robert O'Dette, who authored Synagro's white paper containing allegations of research misconduct against me and my coauthors at UGA. In his reply, O'Dette explained how the system works:[38]

ROBERT O'DETTE, SYNAGRO (2002)

What we don't need are more so-called scientists whose research findings are predetermined by scientific or personal bias. These people will find their work rightly discredited and their funding will disappear while credible researchers continue to have funding.

Synagro sent its own expert, Ian Pepper from the University of Arizona, to give a presentation at our conference, and it held its own conference across the street with others speaking on its behalf.

Department of Justice

In 2004, Andy McElmurray asked a local FBI agent to find out why the Justice Department was prosecuting a South Carolina man for disposing of paint wastes in a sewer line, but not pursuing a City of Augusta employee for intentionally contaminating area farms with biosolids containing illegal hazardous wastes. The agent arranged for Andy and his attorney, Ed Hallman, to meet with the assistant US attorney in Savannah, who said that the case should be investigated, but said that he didn't have the time. They also met with the head of EPA's Criminal Investigations Division in Region IV in Atlanta, who explained that he would not pursue a case involving violations committed by "another government entity."[39]

When I testified in Honolulu, a council member alleged that several of his fellow council members had accepted bribes from someone working for or with Synagro Technologies, Inc. He said he rejected a $5,000 offer, but alleged that more influential members were offered more. Later, he let me look though his files, and I copied several documents that clearly indicated that the bidding process had been rigged by a city official. I forwarded the documents to the head of EPA's Criminal Investigations Division (CID) in Boston, whom I happened to know. After reviewing them, he said that he didn't think the US Attorney in Boston would be interested.

In 2005, the dairy farmers and I filed the qui tam lawsuit over EPA and UGA employees using a federal grant to publish Augusta's fabricated data. We included Honolulu's documents in that case because Synagro was the same company that applied Augusta's biosolids during the Gaskin study and collaborated with Gaskin. The Justice Department, however, was not interested in pursuing the case. Five years later, Mr. Hallman contacted the FBI in Detroit when the Justice Department there prosecuted a city council member for accepting over $60,000 in bribes from Synagro. The contract was reportedly worth $2.1 billion.[40] Hallman forwarded some of my information from Honolulu to the FBI, which confirmed receiving it. They never requested any additional information.

The US Department of Justice, however, doggedly pursued me when EPA asked the assistant US attorney in Macon, Georgia, to collect $61,493 in court costs, plus interest, which the court ordered in May 2012 after dismissing our qui tam lawsuit. These costs covered expenses incurred when private attorneys hired by EPA and UGA copied all of the documents associated with the dairy farmers' previous lawsuits against the City of Augusta. Although none of these documents were used in any of the proceedings, and case law only grants court costs for documents used in the proceedings, the court overruled our objections on the basis that they were untimely.

The assistant US attorney sent me, and only me, an official US Department of Justice form requiring me to produce my personal tax records, and list the current value of all of my assets down to every piece of furniture, including any pictures hanging on my walls. It also required the current balances on any bank accounts I owned, plus any and all other financial information pertaining to my financial assets. Although the form stated that the information was voluntary, it explained that the Justice Department has ways of forcing citizens to provide the information. It included the following admonition in bold print:

WARNING False statements are punishable up to five years imprisonment, a fine of $250,000, or both pursuant to 18 USC. $1001.

In my response to the assistant US attorney, I pointed out that the Department of Justice form states that it only pertains to collecting debts owed to the federal government.[41] The court order stated that the dairy farmers and I owed court costs to the *defendants*, that is, Julia Gaskin and the other UGA and EPA employees who were sued in their private capacity. The False Claims Act explicitly excludes federal and state entities, such as EPA and UGA, from being named as defendants. The court, in fact, rejected EPA's and UGA's arguments when they challenged our ability to sue the defendants in their private capacity.

Moreover, I reminded the assistant US attorney that my attorneys had already provided all of my financial data using the State of Georgia's form for assessing the ability of plaintiffs to pay court costs. The federal government has no corresponding statute concerning court costs in federal cases, and, when that is the case, state laws are applicable.

Finally, in my response to the assistant U.S. attorney, I refused to estimate the current value of my physical possessions on the basis that I am unqualified to accurately determine their current value, and unable to hire the proper experts to assess their value. I explained that, as a board member of the National Whistleblowers Center, I am familiar with cases where government whistleblowers have been prosecuted by the Justice Department and imprisoned over innocent discrepancies in statements made to federal agents. I wasn't about to voluntarily join Bradley Birkenfeld in prison. I did, however, offer to let the Department of Justice have its own assessors come to my home and determine the value of my possessions. The assistant U.S. attorney replied as follows:

ASSISTANT US ATTORNEY TO MR. HALLMAN (2012)[42]

The United States of America entered into an agreement to hire and of course pay private counsel to represent the EPA (Agency of the United States) Defendants in this case. Pursuant to the stated agreement the United States has paid private counsel for representation of the EPA defendants. Therefore, the costs awarded to the EPA defendants in this case are the Government's costs and recoverable by the Government.

In his reply, the assistant US attorney made it clear, once again, that the Justice Department was interested in extracting the government's costs from me, and *only* me, even if it meant forcing me to mortgage my house. At no time did he require either of the dairy farmers to provide their tax records and other financial information. He stated:

ASSISTANT US ATTORNEY TO MR. HALLMAN (2012)[43]

[The] relators' pending offer of settlement in the amount of $35,000.00 will be promptly considered by the Department of Justice upon relator, Lewis', response to the questions below. [Partial List]

- *What is Mr. Lewis' anticipated adjustable income for tax year 2012?*
- *How much equity does Mr. Lewis have in real property and personal property?*
- *What is the estimated value of the real property which is the subject of the pending mortgage or refinance application?*
- *If the above stated mortgage referenced in question#3 is approved how much cash does Mr. Lewis expect to receive at closing or upon approval?*

I informed the assistant US attorney that my house was jointly owned by my wife, and that she was not willing to mortgage her property to pay my debts. Upon reviewing my answers to his questions, he agreed to settle the court costs for $35,000.[44] Before entering into these negotiations, EPA and UGA originally offered to drop the $61,493 in court costs if the dairy farmers and I would agree to a gag order. We refused.

The End

When Speaker of the House Newt Gingrich greeted me in his office overlooking the National Mall in 1996, he looked at me and said, "You know you're going to be fired for this, don't you?"

"I know," I replied, "I just hope to stay out of prison." The speaker had just read my commentary in *Nature*, titled "EPA Science: Casualty of Election Politics."

At first, *Nature*'s editors didn't like this title, which I faxed to London early one morning as I headed off to my job at EPA. Several days later, they let me know that it had grown on them, and they were going with it.

It reflected the proverbial crossroads in my life. Since I was five years old, I wanted to become a scientist and have my own laboratory. A few months before *Nature* published my commentary about the politicization of science at EPA, I finally had my own lab and the freedom to pursue whatever research projects I wanted to within the boundaries of EPA's mission. Henry Longest hung around just long enough to terminate me on my fifty-fifth birthday in 2003.

Several things became crystal-clear to me during the intervening period. First, EPA, UGA, and the Department of Justice had no interest in stopping federal and state employees from using federal grants to publish bogus data to cover up widespread illnesses and deaths associated with EPA's sewage sludge regulations. Secondly, the Justice Department wasn't concerned about Augusta's fabricating twenty years of environmental data required under the Clean Water Act to cover up hazardous wastes in Augusta's biosolids. Finally, each institution was keenly interested in silencing scientists who reported research fraud committed by government employees protecting the government's interests.

As institutions, EPA, UGA, and the Justice Department seemed perfectly happy to let government employees provide the National Academy of Sciences with fake data to conclude that there's no evidence that EPA's sludge rule has ever failed to protect public health and the environment. And the National Academy of Sciences had no interest in correcting the record after *Nature* published an editorial and news article about problems with the Gaskin study. The Pennock family in Pennsylvania, who lost their son Danny to biosolids, got together with Andy McElmurray and filed a formal complaint against Tom Burke, the chair of the NRC panel, for removing the one last shred of evidence from the NAS report of what really happened in these cases.

They were part of a small group of families who lost their children and livelihoods who made the national news. EPA assistant administrator Tracy Mehan swept them under the rug in just one letter, citing the National Academy of Sciences report, the Gaskin study, and a medical examiner's preliminary report in New Hampshire, on Christmas Eve in 2003. The examiner's report, which dismissed any link between biosolids and Shayne Conner's death, was based on literature produced under the EPA-WEF National Biosolids Public Acceptance Campaign, which was provided by the NH Department of Environmental Services (DES):

ACTING CHIEF MEDICAL EXAMINER, NH (1995)[45]

*After review of pertinent literature, it is my understanding that
this type of fertilizer has been tested for a number of hazardous
parameters, and found safe to use. There appears to be no scientific basis for connecting this person's sudden and tragic death to
any environmental or infectious hazards posed by the use of such
material.*

This opinion, however, was contradicted by DNA analyses that my
research team performed on bacteria present in frozen samples of the biosolids collected in Shayne Conner's neighborhood at the time of his death.
The samples were overgrown with *Brevibacterium diminuta*, a type of bacteria
known to cause sudden respiratory failure and death when inhaled with dust
particles. Its presence could easily explain why residents inhaling biosolids
dusts blowing from the field treated with biosolids near their homes experienced difficulty breathing, and Shayne died of sudden respiratory failure. The
Complaint eventually settled out of court.

EPA's assistant administrator took care of them all and my life's work
as well. He wrapped them up as a Christmas gift, and gave it to science and
industry, the children of liberty. There were, however, many more who suffered and died. I visited some of their families in their homes, and read their
desperate letters and emails that keep pouring in year after year. They are
cropping up in every major city and rural area across America, looking for
someone, anyone who can help. With EPA and the biosolids industry suppressing independent investigations into adverse health effects, those who
are harmed can no longer find expert witnesses willing or able to testify on
their behalf. It's the government's way of protecting the industry, and itself,
against prosecution.

Other farmers in Georgia and elsewhere are experiencing what happened on the dairy farms near Augusta.[46] But, after seeing what EPA did
to the McElmurray and Boyce families, most farmers keep their silence.
A reporter in Mobile, Alabama, called me about several dairy farms there
having the same problems with biosolids, but wouldn't report it for the same
reason. One former executive with a national company in the biosolids business told me off the record that cities across America make up data on pollutant levels in biosolids, not just Augusta. He said everyone knows that EPA
will cover them, just like it did Augusta.

This is not an isolated case. It's how government, industry, and our academic institutions use science to protect government policies and industry practices. Any scientists who stand in their way will have the same outcome as I did. Any private citizen who tries to fix the system will get the same treatment as the McElmurray and Boyce families did. On my last day at EPA, there was no retirement party. I drove over to the BBQ Shack, a popular hangout for EPA employees, for lunch. Dr. Larry Burns, who was an associate director for the National Exposure Research Laboratory in Research Triangle Park, North Carolina, came over to my table to say goodbye. He served as a liaison between our lab and the national lab in Research Triangle Park. "You should be proud that the National Academy of Sciences supported your work," he said.

"I just hope somebody at EPA continues it," I replied.

"No way," he said, "not after what happened to you."

Later that day as I headed out the door of the EPA lab on College Station Road, I stopped by the research director's office to say goodbye to Harvey Holm. I had begun working there as a Stay-in-School student in October of 1968, fresh out of high school. "I can't believe you're really leaving," he said. "I thought it would never happen. You'll be back, I'm sure of it." No, Harvey. I won't ever be coming back.

In cases such as mine, the ultimate objective of government and industry is to shut down debate. Recently, a reporter told me that she spoke with other scientists about research misconduct supported at institutional levels in order to increase the flow of grants from government and industry. All of them said that they've never been pressured by their employers to support government or industry. I suggested she go back and ask them what they think would happen if scientists at land grant universities, which are heavily funded by the USDA, published data that undermined USDA policies, or what the ramifications would be if scientists working at medical schools heavily funded by the pharmaceutical industry linked vaccines to autism or Alzheimer's Disease.

Herein lies the biggest problem of all. The damage to scientific progress caused by skewed data is likely to be relatively small compared with the cumulative impact of the research that scientists avoid publishing to protect their careers. In other words, the greatest threat to scientific integrity isn't what gets published; it's the research that doesn't get published—or even conducted in the first place. The ultimate goal of EPA's National Biosolids Public Acceptance Campaign, according to the agency's decision memorandum, was to make biosolids "non-controversial by the year 2000." If scientific debate were unfettered, the scientific community would, sooner or later, determine which side of an issue is correct. Leaders of government and industry know this; hence, their goal is to silence debate in areas of science where their interests are at stake.

BLACK MAGIC

BIOSOLIDS: DISPROPORTIONATE
IMPACT ON BLACK COMMUNITIES

In a 2004 study of land application of biosolids published in the National Institutes of Health journal, *Environmental Health Perspectives*, David Gattie, a UGA professor of engineering, and I wrote:[1]

> *One outcome of local bans is that land application of sewage sludge is being forced out of areas where residents have the political and economic resources to oppose the practice and into economically depressed areas. Whether this is intentional or not, sewage sludge is being dumped more and more into those communities least able to have their complaints heard, and where residents are least capable of relocating or obtaining medical treatment.*

Currently, we are assisting Billups Grove Baptist Church in Athens, Georgia, where members of its congregation are suffering adverse health effects from the city's biosolids composting operation close to the church (see Appendix III: Some Things Never Change). This historic church, which was organized in 1898 by former slaves, holds an annual march to the Athens–Clarke County landfill and biosolids operation on the first or second Saturday of May. In 2013, the march was replaced by an outdoor event at the Valley of Billups Grove, which we built in Oconee County with funds donated by the Segal Family Foundation and Focus Autism, Inc. The Valley provides the Billups Grove community with a place for holding outdoor events away from the stench of biosolids.

In 2013, Steve Wing, Amy Lowman, and others at the University of North Carolina (UNC) followed up with a survey of thirty-four individuals, ages thirty-five through eighty-three, living near biosolids land application sites in North Carolina, South Carolina, and Virginia (twenty-one white, twelve African American, and one Hispanic).[2] Nearly all of the residents reported offensive odors, which some described as "unbearable." Others "got used to it," and one reported, "It don't bother me." The odors lasted from two days to six months after application. Consistent with what we reported in 2002, over half of the residents complained of eye, nose, and throat irritations and gastrointestinal symptoms (nausea, vomiting, diarrhea).[3] Other symptoms included coughing, difficulty breathing, sinus congestion or drainage, and skin infections or sores.

About one-third of residents in Lowman's study noted changes in the natural environment. Seven reported more deaths and illness among livestock and water life. For example, one resident reported:

> I look at the sludge on this slope—when they put it out, if it rains, this water flows down in this branch.... Now there is no fish or anything that lives in these little branches. No crawdads, anything.... When I was growing up, we'd go there and I would fish for them and so forth. But all this is gone.... So that is saying something has killed all this stuff.

Five reported a change in private well water, including "green slime" and odor. For example, one man said:

> My well...water had an awful smell to it, and a green slime... like three months [after sludge application].... Before they [applied sludge], I had lived here...two and a half years. Without a problem.

When the study was published, Ned Beecher, director of the Northeast Biosolids & Residuals Association (NEBRA), requested that the National Institute of Environmental Health Sciences (NIEHS), which funded the study, investigate Prof. Wing for research fraud.[4] Beecher and others discussed their efforts to Wing and his research in an "ABBA Quarterly Conference Call," which Beecher facilitates for EPA, Synagro and others who promote biosolids and discredit scientists and local residents who report adverse health effects of treated sewage sludges.[5]

> *NBMA is now not sure if they want to keep bringing attention to the report or just let it go. The funding supporting this research may not be available anymore... Steve Wing is 58 and is not a tenured professor at Chapel Hill.... There will not be much more happening with this other than in Chapel Hill. Not sure if we are going to take it to another level. Enough may have been done.*

Prof. Wing initially participated in efforts by the Water Environment Research Foundation (WERF) to investigate reports of adverse health effects linked to biosolids. He later dissociated with WERF and published an article about the organization's conflicts of interests and strings attached to its funding.[6] ABBA is a leading manufacturer of industrial pumps used by wastewater treatment facilities. Twenty-four individuals participated in the conference call, representing Synagro Technologies, Inc., the Water Environment Federation (WEF), National Biosolids Partnership (NBP), and others. The NBP is operated by the WEF in collaboration with the National Association of Clean Water Agencies (NACWA) and local and regional biosolids management organizations across the United States and Canada with support from the US Environmental Protection Agency. [7]

Beecher and his organization, which are funded by EPA and Synagro Technologies, Inc., are the same group that published Synagro's false allegations of research fraud against me and my coauthors in 2001, then again in 2012 when I published an article about institutional research misconduct associated with EPA's biosolids program.[8]

In 2003, when our study highlighting environmental justice concerns over EPA's biosolids program was published, the UGA engineering department prepared a press release to go with it. In the package, I included information about an impoverished African American community that I investigated in Grand Bay, Alabama.[9] Whenever the fields were treated with biosolids, children there who drank water from wells in or near the fields experienced severe gastrointestinal problems, and experienced painful cramps in their legs that prevented them from walking to school for weeks at a time. They also had severe respiratory problems from inhaling dust blowing from the treated fields, which spread out in all directions from their homes and reached as far as the eye could see. Their parents and grandparents stayed cooped up in their houses with no air conditioning during oppressively hot summers, and breathed through rags to filter the dust.

UGA, however, quashed the press release. Its Scientific Integrity Officer, Dr. Regina Smith, later testified:[10]

> *I read the article, and I read the press release, and [UGA VP for Research] Gordhan Patel and I talked, I think, and my conclusions were just because you publish an article in some little-known journal is no reason to put out a press release. You do press releases when you're publishing in important journals, like* Nature, Science, *and* Cell. *Not in something called* Environmental Health Perspectives.

As mentioned in the previous chapter, UGA issued a national press release on an EPA-funded project published in the *Journal of Environmental Quality* earlier that year, which concluded that biosolids implicated in killing dairy cattle on two Georgia farms *did not pose a risk to animal health*. Lead author Julia Gaskin, a land-application specialist with a bachelor's degree in psychology and a master's in forest hydrology, was quoted in the press release: "Some individuals have questioned whether the 503 regulations are protective of the public and the environment. This study puts some of those fears to rest."[11] Regina Smith testified that this press release, titled "Sludge study relieves environmental fears," was not brought to her attention.

As I did with the African American community in Grand Bay, I investigated a similar case in Convent, Louisiana, where EPA and state public health officials dismissed health problems reported by 185 residents in a predominately African American community. There, residents began complaining of burning skin, boils, and rashes when the City of Kenner spread its biosolids in fields of sugar cane from 2000 through 2004. Health officials identified *S. aureus* as causing skin and eye infections, but ruled out biosolids as playing any role. Instead, they blamed poor personal hygiene and addressed the complaints by holding a "health fair" to instruct residents in proper bathing and cleanliness. EPA and state health officials did not assess the personal hygiene habits of affected versus unaffected residents, and did not acknowledge any of our peer-reviewed research, which linked numerous cases of *S. aureus* infections to chemical irritation caused by biosolids.[12]

EPA Response: Biosolids Prevents Lead Poisoning

What is the relationship between environmental justice and childhood lead poisoning?

Environmental justice says that no group of people should bear an uneven burden of harmful environmental results. Environmental injustice occurs when environmental effects have a harsh impact on minority and/or poor populations that is greater than the harsh impact on the general population....

Progress has been made in reducing children's blood-lead levels in the United States. However, average blood-lead levels remain unequally high among non-Hispanic Black children when compared to Mexican-American and non-Hispanic White children. This is an example of a disproportionate disparity. Risk factors for higher lead levels include older housing, poverty, and being non-Hispanic Black. Many children with high blood lead levels are also unequally affected by other environmental issues in their communities.

CDC National Center for Environmental Health

In 2000, scientists at the USDA and Johns Hopkins' Kennedy Krieger Institute (KKI) applied biosolids containing lead and other hazardous wastes to an African American neighborhood in Baltimore, Maryland. The purpose of the experiment was to see whether biosolids containing high levels of both iron and phosphorus would render potentially dangerous levels of lead in the contaminated soil non-bioavailable, that is, incapable of being absorbed by the digestive tract in children who consumed the biosolids and soil. Small children often directly ingest lead-contaminated soils when playing in their yards, and when the contaminated soils are tracked into their homes. They also inhale dusts, and pick up dusts from handling contaminated surfaces.

The Baltimore study, which was funded by EPA, USDA, and the Department of Housing and Urban Development (HUD), was published in *The Science of The Total Environment* in 2005.[13] It was headed by Dr. Mark Farfel at KKI and Dr. Rufus Chaney.

In 2001 EPA and the USDA also experimented in a poor, black neighborhood in East St. Louis, Illinois.[14] Based on these and other studies with humans and animals, they concluded that tilling "Class A" biosolids into lead-contaminated soils in inner-city neighborhoods is advisable to reduce the risks of lead poisoning. Class A refers to sewage sludges treated with low levels of lime, heat, or chemical processes capable of reducing indicator pathogens to undetectable levels. But the indicator pathogens recommended by EPA, for example, *Salmonella*, are highly sensitive to heat and chemical disinfection. Most viruses and bacteria found in sewage sludges can survive these processes.[15]

I was questioned about Chaney's lead containment theory in court proceedings held by the US Department of Labor in 2003.[16] Prior to my testimony, Dr. John Walker, EPA's national spokesperson for biosolids, testified that Chaney's theory is supported by a number of studies funded by EPA and the USDA. As mentioned in the previous chapter, most research scientists in EPA's Office of Research & Development (ORD) considered the theory that biosolids render lead and other pollutants non-bioavailable to be bogus, and commonly referred to it as "sludge magic."

DIRECT EXAMINATION BY MR. KOHN:[17]

Q. Dr. Lewis, do you remember testimony from Dr. Walker regarding whether Class A sludge, if children ate it, it may help them in terms of lead poisoning?

A. I do.

Q. And do you understand the science behind ingesting Class A sludge and its relationship to heavy metals like lead?

A. I would say I understand the science. I understand the theory. There's no science that supports that in my opinion.

Q. And is Dr. Walker's position on that accurate from a scientific perspective?

A. No, it is not.

Q. Can you explain why not?

A. That issue goes back to the genesis of what became known within the Office of Research and Development as "sludge magic."

Basically that the theory is that heavy metals such as lead become tied up or bound with the organic fraction that is in sewage sludge so that they're not available to be taken up by humans or animals or plants. They are perpetually and forever bound.... First of all, the human stomach, which has hydrochloric acid in it, has an extremely low pH, which would [dissolve the] heavy metal....

JUDGE TURECK: *I'm very glad to hear you say that it has no scientific basis because the thought of mothers of America going around feeding their children sludge is not a pretty one.*

In 2007, I decided to share some of my documents on the Baltimore study with John Heilprin, an environmental reporter working for the Associated Press in Washington, DC. After working with me for about a year, he and Kevin Vineys put together a five-part series on biosolids, beginning with the lead experiments in Baltimore.[18] An expert on biosolids at Cornell University, Dr. Murray McBride, commented:

In a newsletter, the EPA-funded Community Environmental Resource Program assured local residents it was all safe. "Though the lot will be closed off to the public, if people—particularly children—get some of the lead contaminated dirt in their mouths, the lead will just pass through their bodies and not be absorbed," the newsletter said. "Without this iron-phosphorus mix, lead poisoning would occur."

Soil chemist Murray McBride, director of the Cornell Waste Management Institute, said he doesn't doubt that sludge can bind lead in soil. But when eaten, "It's not at all clear that the sludge binding the lead will be preserved in the acidity of the stomach," he said. "Actually thinking about a child ingesting this, there's a very good chance that it's not safe."

EPA and USDA apparently had no qualms about testing biosolids containing hazardous wastes on humans. I've seen records documenting experiments funded by the USDA in which prisoners were fed capsules of biosolids added to soil from a superfund site. In a deposition conducted by one of my attorneys, Ed Hallman, Rufus Chaney discussed human trials that the USDA carried out at Columbia University, but denied that the human subjects were prisoners:[19]

DEPOSITION OF DR. RUFUS CHANEY BY
MR. HALLMAN:

A. You know we even did a human feeding test to show that we could treat soil with phosphate and reduce the bioavailability.

Q. Is that something about pills given to prisoners or something?

A. No.

Q. Where did you feed people?

A. At Columbia University a test was conducted with all the proper approvals, feeding soils from the Joplin, Missouri, experiment using biosolids compost, phosphate, and other soil amendments to attempt to reduce the bioavailability of lead in a smeltered polluted soil.

Heilprin told me that, as soon as the Baltimore story ran, Johns Hopkins University pressured AP to back off.[20] AP, therefore, never published the rest of the series. Rufus Chaney emailed Julia Gaskin and UGA Dean Jay Scott Angle, Chaney's former collaborator at the University of Maryland, claiming that AP punished Heilprin for misstating the facts, and planned to publish a retraction.[21] One of my attorneys, Ed Hallman, however, emailed AP senior managing editor Mike Silverman, who denied Chaney's claims.[22] In his deposition, Chaney explained what transpired between Johns Hopkins University and AP.[23] At first, he stuck by his story that Heilprin was disciplined by AP. Then, after being confronted with Silverman's email to Mr. Hallman, Chaney denied saying that Heilprin was disciplined:[24]

DEPOSITION OF DR. RUFUS CHANEY BY
MR. HALLMAN:

Q. And how do you know—who is the reporter, is it John Heilprin?

A. That's the one.

Q. How do you know he was disciplined?

A. Because I was contacted by the AP reporter who wrote the subsequent article and by an AP manager to discuss the case.

Q. And they told you that he had been disciplined?

A. What do you think happens when you're assigned—when your assignment is shifted? Mr. Heilprin had moved from the United Nations to this wonderful new opportunity of environmental reporting for AP. He was re-assigned and they apologized for his mistaken articles. Now you tell me if that's not being disciplined....

Q. When you were complaining about Mr. Heilprin, did you talk directly to the AP folks?

A. I was called by the AP manager and the AP reporter who did the follow-up article.

Q. And did you do any—did you have any conversation with anyone at Johns Hopkins about the article?

A. I had several conversations with managers at Johns Hopkins about their response to the Heilprin articles.

Q. And who were they?

A I don't recollect names. They were officials of the research corporation and the university.

Q. And did they complain to AP, too?

A. It was their complaint to AP that led to AP actually doing something....

Q. I'll ask you to look at Exhibit 239 and ask you if you've ever seen this document before? Have you ever seen this before?

A. I haven't seen Hallman's, no....

Q. And this document says: Dear Mr. Hallman: I am writing in response to a request for information sent to John Heilprin by Kathryn Sims in your name. This e-mail will confirm that Associated Press reporter John Heilprin was neither punished nor disciplined in any way for his stories examining the safety issues involved in the use of sewage sludge. Please confirm receipt.

I believe you say that you got something in writing stating that he was disciplined; is that correct?

A. I didn't say what you just said I said.

Q. You didn't say in writing?

A. I said that he was re-assigned, is no longer the lead environmental beat reporter for AP, and they subsequently published what everybody else interprets as a retraction of most of the claims of the Heilprin article at least about Baltimore.

Although Heilprin told me that AP backtracked on the Baltimore story and shelved the rest of his series because of pressure from Johns Hopkins' lawyers, Chaney had the details concerning Heilprin's situation at AP backwards. Just as the first installment of his biosolids series went to press, AP offered Heilprin a plum position as a UN correspondent. In his new job, he covered the news worldwide. That same year, the United Nations Correspondent Association awarded him the 2008 Elizabeth Neuffer Memorial Prize, a gold medal, for "his coverage of Myanmar and adroit use of his trip with the Secretary Media General to get in."[25]

A couple of months after the Baltimore story ran, the Office of Civil Rights for the Attorney General in Baltimore asked me to review the USDA study.[26] It reported that tilling the soil raised the minimum soil lead concentration from 479 ppm to 800 ppm, and adding composted sewage sludge further increased the lead levels. The resulting lead levels were high enough to cause critical brain tissue atrophy in areas of the brain controlling executive functions, mood regulation, and decision making, and may contribute to criminal behavior.[27] EPA's Final Rule 66, which governs concentrations of lead permitted in children's playgrounds, sets the maximum safe level at 400 parts per million (ppm) of lead for bare soil in play areas.[28]

By EPA's own standards, therefore, tilling the soil and adding lead-contaminated biosolids could have potentially increased the risks of lead poisoning in children to dangerous levels. Even though UDSA believes that the biosolids made the lead unavailable, the fact remains that Chaney and his collaborators at Johns Hopkins and elsewhere chose to experiment with African American children, who were placed at risk of brain damage if the experiments failed.[29]

In a study assessing the lead abatement benefits of landscape coverings, researchers in the Department of Pediatrics at Northwestern University in Chicago found that grass coverage of bare soil areas cut the amount of track-in of lead-contaminated dusts in half.[30] Citing an EPA risk assessment, the authors concluded:[31]

> *Lowering the track-in of lead may be the most important influence for the youngest children, whose main pathway for blood lead elevation is lead-contaminated floor dust adhering to a hand, which is then placed in the mouth.*

In other words, when the Baltimore study was done, EPA was aware that the main pathway for elevating blood lead levels was via contaminated soil dusts tracked in and deposited on floors where small children crawling on hands and knees would transfer the dust to their mouths. Despite knowing this, EPA funded Farfel and coworkers to till the contaminated soil, raising both soil lead concentrations and dust levels. Farfel and coworkers collected dust samples but, unlike researchers at Northwestern University, they only reported the relative "bioavailability" of the lead in the contaminated dusts. They did not report the total amounts of lead-contaminated dusts that accumulated in main entryways or on doorway mats provided to the study participants in Baltimore.

Moreover, the authors' conclusions about bioavailability were based on the assumption that bioavailable forms of lead were transformed to a mineral called pyromorphite, which renders them essentially permanently non-bioavailable. Yet, they never analyzed any of the soil samples in Baltimore to see whether pyromorphite was even present. Worse yet, in a previous study on pyromorphite authored by Rufus Chaney and others, they wrote:[32]

> *[We] will only be able to quantify and predict these observed changes in lead bioavailability when the mechanisms that affect lead solubility and absorption by the receptors are understood.... Moreover, without an understanding of the reasons for variations, a change in measured lead bioavailability is of limited value.*

Farfel and coworkers, simply put, had no way of knowing whether contaminated soils in the Baltimore study would be rendered safer by virtue of lead being transformed to pyromorphite. Lead bioavailability in the neighborhoods treated in Baltimore, of course, could be reexamined today to determine whether the effects that Farfel and coworkers reported have lasted as they claimed they would.

Note: Native American residents living on reservations are treated similarly to African American communities. See Appendix IV.

UGA and Republicans Torpedo Senate Hearings

The day after AP published the Baltimore story, Heilprin and Vineys ran a breaking story on Senate hearings.[33] The article began:

> *The Senate Environment and Public Works Committee will investigate the government's funding of research in poor, black neighborhoods on whether sewage sludge might combat lead poisoning in children, its chairman said Monday.*

> *The Associated Press reported Sunday that the mix of human and industrial wastes from sewage treatment plants was spread on the lawns of nine low-income families in Baltimore and a vacant lot next to an elementary school in East St. Louis, Ill., to test whether lead in the soil from chipped paint and car exhausts would bind to it.*

Chairman Barbara Boxer invited Andy McElmurray, one of the dairy farmers affected by Augusta's biosolids, and me to testify.[34] A staff member working for Senator James Inhofe, the ranking minority member on Boxer's committee, also contacted me about testifying. Andy and I accepted Boxer's invitation, and met with her staff. Republicans protested the hearing. It was eventually downgraded to a briefing, which both EPA and Republican members of the committee vowed to boycott.

Mr. Hallman represented me and the dairy farmers in two qui tam lawsuits involving Augusta's biosolids. Qui tam lawsuits allow private citizens who uncover fraud against the US government to sue on behalf of the government under the False Claims Act. The US Treasury keeps all but a small portion of the funds, which goes to the plaintiffs, called *relators*. We filed our first qui tam lawsuit against the City of Augusta, Georgia, for false claims it made when borrowing tens of millions of dollars from a government program established under the Clean Water Act to help wastewater treatment plants maintain their facilities.

After the dairy farmers filed their first lawsuits against Augusta in 1998, the Georgia Environmental Protection Division (EPD) investigated the city's wastewater treatment plant. EPD found it was "in shambles," and recommended that its land-application program be shut down immediately.[35] The inspectors confirmed what experts hired by the dairy farmers had already proven, namely, that Augusta's environmental monitoring data reported to EPA were completely unreliable. In the process, another audit revealed that Augusta had used the government funds it borrowed to repair its wastewater treatment plant to build a river boardwalk to attract tourists to the city. Unfortunately, neither EPA nor the EPD were interested in getting the US Department of Justice to prosecute the city and recover any of the funds. So, the dairy farmers and I filed a qui tam lawsuit against Augusta based on the fact that we were an original and independent source of information proving that the City of Augusta knew that the information it included in its grant applications was false.

My investigations documented human illnesses linked to Augusta's biosolids, and some of the environmental samples I collected indicated that Augusta's biosolids were saturated with toxic chemicals. For example, a county worker contacted me after developing severe respiratory problems from spreading hay treated with Augusta's biosolids.[36] One of his coworkers was even rushed to the emergency room with difficulty breathing after spreading a few bales of the hay. Jennifer Lee, who was reporting on the story for the *New York Times*, accompanied me late one night when one of the road workers showed me some bales of hay stacked for workers to spread the next day.[37] She went over to look at them while I was still gathering some equipment, and came back saying that they had a horrible smell that made her feel light-headed. I could smell it too, even before getting close enough to handle it. It had an overwhelming odor of pyridine, a chemical solvent used to manufacture agrochemicals and pharmaceuticals.

The results of independent investigations were consistent with what EPD knew, but was unwilling to act upon. Records, which the dairy farmers and I obtained under the Georgia Open Records Act, showed that Augusta's wastewater treatment plant had a long history of failing to enforce pretreatment regulations. Under the Clean Water Act, pretreatment regulations are designed to reduce the toxicity of chemical wastes before they are discharged into sewer systems. But the EPD found that Augusta was allowing Coca-Cola to discharge "off-spec" material directly into the sewer system without dilution or any pretreatment. NutraSweet was also discharging hazardous chemical wastes into the sewer system without recording the amounts. Castleberry Foods Company and Amoco were "chronically non-compliant." EPD described Castleberry's pretreatment system as "extremely degraded." Blackman Uhler, a chemical manufacturer, was discharging methyl chloride into the sewer system in concentrations as high as sixteen times the legal maximum concentrations. It also discharged illegal amounts of phenol and nickel into the sewer system "on a consistent basis."

We filed the second qui tam lawsuit over a $12,000 grant EPA gave UGA in 1999.[38] The defendants in our lawsuit were EPA employees, including Dr. John Walker and others who arranged the grant, and UGA employees, including Julia Gaskin and others who administered it. This grant was used by EPA and UGA to publish twenty years of environmental monitoring data obtained from the City of Augusta, Georgia, which were required under the Clean Water Act. The data indicated most regulated heavy metals in Augusta's sewage sludge dropped to safe levels when EPA passed the 503 sludge rule in 1993.

A federal court in a different lawsuit filed by Andy McElmurray ruled that all of these data were fabricated by the City of Augusta to cover up cattle deaths caused by the city's sewage sludge. Furthermore, based on the kind of information I described above in our first qui tam case, the court ruled that Augusta's environmental monitoring data were unreliable. And the court made it clear that the reason it was common knowledge is because experts hired by the McElmurray and Boyce families determined that heavy metals and other hazardous wastes were present in Augusta's sludge at much higher concentrations than the city was reporting to the State of Georgia. In handwritten notes in the margin of the paper Julia Gaskin submitted for publication, one of her coauthors, Bill Miller, scribbled the following note next to conclusions based on Augusta's data: "We should fess up here that we don't know exact rates of application or specific characteristics of the sludges applied."[39]

In our second qui tam lawsuit, therefore, we alleged that the defendants included various false statements in their grant application concerning the purpose of the grant and the quality of data that would be produced. Our evidence proved that its purpose, from the beginning, was to protect EPA's regulations and policies on biosolids. As stated earlier, Gaskin herself claimed in a national press release issued by UGA, "Some individuals have questioned whether the 503 regulations are protective of the public and the environment. This study puts some of those fears to rest."[40] Under the Federal Grants and Cooperative Agreement Act of 1977, it is illegal to use federal grants to support federal policies and regulations, and all federal grant applications require recipients to affirm that this is not their intent. Because the National Academy of Sciences used the Gaskin paper to dismiss the cattle deaths and conclude that there is no documented evidence that EPA's 503 sludge rule has failed to protect public health and the environment, we believed it was important to file this lawsuit even though the amount of the grant was small.

Before filing a qui tam lawsuit over the Gaskin study, I contacted UGA's legal counsel, Arthur Leed, and offered to drop the matter if Gaskin would withdraw the paper because some of the data were fabricated.[41] UGA refused.[42] I also had my attorney, Ed Hallman, send Gaskin and her coauthors documents proving that Augusta fabricated the data they had published, and asked them to withdraw the paper. Gaskin discussed the letter with UGA's scientific integrity officer, Dr. Regina Smith, who informed provost Arnett Mace, and the VP for research, Gordhan Patel, that Gaskin had no knowledge that the data were fabricated when the paper was submitted.[43] Gaskin, in turn, informed Mr. Hallman that UGA had cleared her of any wrongdoing, and claimed that there was no mechanism for withdrawing the paper.[44]

After agreeing to testify, I decided to inform UGA about the hearings and offer the university one last opportunity to withdraw the paper. In turn, we would drop the lawsuit. Of course, to do this, Hallman's law firm, the dairy farmers, and I would have to absorb all of his costs associated with our lawsuit, which were considerable. We would also have to forget ever being compensated for any damages EPA and UGA caused by using Augusta's fake data to discredit the dairy farmers' legal claims and my research. Otherwise, it would look like we were trying to use the Senate hearing to basically bribe UGA to settle the case. By that time, EPA and Senator Inhofe had informed Boxer that they would boycott the hearing. So Boxer downgraded it to just a briefing.

After I informed our attorneys, Mr. Hallman emailed me a confidential settlement proposal. He too was willing to forfeit everything his law firm had invested in the case if Gaskin would withdraw the paper. But he couldn't stand the thought of the dairy farmers and me walking away empty-handed. He included $100,000 for each of the dairy farmers and a temporary job for me. I drove over to Atlanta and expressed my concerns about how the letter sounded, but let it go. I should have done more to stop it.

UGA released the settlement offer to the Georgia Attorney General's Office, and UGA and/or the attorney general released it to the public media. Boxer's chief of staff called me and Andy in our hotel rooms the night before the briefing to let us know that the hearing had been canceled. AP quoted a spokeswoman for the committee, saying that the hearing was canceled "out of concern that the private litigation would distract from the main issue of sludge safety."[45] The AP article quoted Hallman's letter, stating, "The pair (Lewis and McElmurray) warned that they would be testifying at the hearing and said a settlement would allow them to 'praise UGA for its handling of this matter.'" The Court Record explains what happened:

> *[Plaintiff ("Relator") David Lewis] met with William Boyce, Andy McElmurray and Ed Hallman on or about August 14, 2008, to discuss Relator Lewis' interest in settling the qui tam lawsuit. Several days prior, Relator Lewis telephoned Mr. Hallman and the dairy farmers to inform them of his desire to settle the case before testifying at a Briefing before the US Senate Environment Public Works Committee. Mr. Hallman suggested that the Relators meet with Mr. Hallman at Mr. Boyce's home in Keysville to discuss the matter.*

> *As the Relators sat down, Mr. Boyce put his hand on one of the chairs in his living room and said: "Dr. David, this is the chair Ms. Gaskin was sitting in when she told me that my land was too contaminated for growing food-chain crops." Relator Lewis states that he looked at the pictures on the wall of the Boyces' two teenaged boys standing next to their dairy cows with blue ribbons. He stared at the picture of their son who was killed early one morning in an automobile accident as he drove to check on his dying cow....*

> *Realtor Lewis said that he fought back tears as he contemplated what he was about to ask Mr. Boyce to do. The Relators talked about*

all that they had lost as a result of the Defendants promoting sewage sludge and attempting to quash any questioning of the EPA's bio-solids program. Two of Georgia's most productive dairy farms were destroyed, along with a way of life for the McElmurray and Boyce families. Dr. Lewis' career as a scientist was gone. They spoke of the millions of dollars in resources that their attorneys have invested in prosecuting their cases over the past 15 years.

Relator Lewis, however, explained that the qui tam lawsuit was about the Defendants lying on grant applications and fabricating scientific data to defraud the United States Government. He urged Mr. Boyce and Mr. McElmurray to make a onetime offer to the University of Georgia Defendants that they simply agree with Judge Alaimo's ruling concerning the fabricated data that the EPA Defendants provided to UGA. Dr. Lewis suggested that Relators Boyce and McElmurray give the UGA Defendants the benefit of the doubt and not ask for a monetary settlement. Both agreed to do so.

Purely for the public good, therefore, the Relators were willing to set aside all that they had lost. They were willing to let Ms. Gaskin and the other UGA Defendants settle with their careers and reputations intact and have all of the things that the Realtors could never regain for themselves. Moreover, Relator Lewis [stated] that although Mr. Hallman could not bring himself to propose a settlement in which his clients would receive no money, he asked for no attorneys' fees or expenses for himself or any of his associates.

According to Dr. Lewis, the Relators were hopeful that the University of Georgia would seize this opportunity to correct the scientific record for the public good and not engage in any protracted and costly litigation. They thought that UGA would probably agree to Ms. Gaskin submitting an erratum to correct the scientific record if the Relators would agree to no monetary payments. Dr. Lewis states that the Relators would have accepted such terms, obviously, because they had already agreed among themselves to do just that.

I was also questioned by UGA's attorneys about the settlement offer during depositions:[47]

UGA ATTORNEY QUESTIONS LEWIS (2009)

Q. Okay. Do you recall the September 3rd settlement offer?

A. In this case?

Q. Yes.

A. Yes.

Q. All right. Do you remember suggesting that you would praise one or more of the defendants if some statement were signed?

A. I remember that language in Mr. Hallman's letter.

Q. What did you mean by praise?

A. It wasn't my letter. I don't know what he meant.

The statements that Andy McElmurray and I prepared for the Senate briefing appear in Appendices V and VI.

HOLY SCIENCE

R ecently, the Church of England criticized environmental activists who protested the church's position on the safety of fracking. Philip Fletcher, who advises the General Synod and the archbishops of Canterbury and York, was quoted in *The Telegraph*:[1]

There is a real danger of distorting the arguments through protest. If we take an example from a little way back you will recall the completely misguided attack on the MMR jab. [That] is still causing outbreaks of measles now in our vulnerable children because it gave many households and many parents the wrong impression that MMR was somehow dangerous. You could be in the same position with fracking if you just take the very limited views being expressed by some opponents of fracking at the moment.... I don't think that is at all sound scientifically.

Since the White House began funding faith-based initiatives under President George W. Bush, religious organizations have become an increasingly important mouthpiece for parroting the products of institutional research misconduct designed to support government policies and industry practices. Their strategy, as the example above illustrates, is to accuse scientists and community activists who raise concerns of putting public health at risk. It makes a lot of sense. Who could government and industry choose to carry that message with more moral authority than religious institutions?

Religious Leaders Rising

Government and industry, increasingly, are involving religious organizations in their efforts to support government policies and industry practices. In 2011, for example, the CDC and Department of Health and Human Services (HHS) held an "off-the-record, not-for-press-purposes" phone conference with church and community leaders to administer vaccines in churches, synagogues, and mosques.[2] Under this program, religious leaders are encouraged to have their congregations vaccinated by Walgreens and other pharmacies. HHS writes:[3]

> *As you know, faith and community leaders play an integral role in helping to keep their communities and congregations healthy, especially during flu season. As trusted messengers, you are able to spread important information about healthy practices and the need for vaccination.*

Similarly, at a recent conference of the National Institute of Environmental Health Sciences, local activists combating environmental injustices associated with Concentrated Animal Feeding Operations (CAFOs) talked about the negative impact religious leaders are having.[4] As part of their strategy to control public opposition, companies running the CAFOs created nonprofit organizations and included locally influential clergy on each of their boards.

With both science and religion, the intrinsic value of their institutions rests wholly upon integrity. People can live with car dealerships that unload a lemon or two on unwary customers, and fruit stands that sell a few wormy apples from time to time. But science lacking the highest standards of integrity is worthless. It's like a preacher who only runs around on his wife once in awhile. Scientific organizations cannot afford to sanction unethical behavior by individuals acting on behalf of their institutions, while punishing those who do the same things to serve their own self-interests. Hypocrisy knows no bounds. As institutions become more corrupt, they are more prone to punish the innocent. They do so, at least partly, to create an appearance that they don't condone the very acts that they themselves commit with reckless impunity. According to Robert Kuehn:[5]

> *There is a long history of attacks on scientists. During the Inquisition, the Roman Catholic Church charged Galileo with*

*heresy and, after imprisonment and threats of torture, forced
him to renounce his theory that the sun, not the earth, was the
center of the universe. In the 1950s, politicians sought to silence
scientists that allegedly held political views sympathetic to
Communists. In recent years, research results, rather than the scientist's religion or politics, have motivated attacks on scientists.*

Kuehn's snapshot of the motives behind attacks on scientists over the
past four centuries seems to suggest that religion no longer plays an important role. The kinds of conflicts between science and religious fundamentalists that cost Galileo his freedom, however, rage on to this day. They
determine what public schools are allowed to teach about evolution, and set
the boundaries of stem-cell research and the causes of climate change. As
incredible as it may seem, religion played an important role in the attacks on
me by Henry Longest, EPA's deputy assistant administrator for management
for the Office of Research & Development (ORD).

Turning to One Another at EPA

In 2003, Longest held leadership meetings to develop ORD's next
generation of managers. For guidance, he distributed Margaret Wheatley's
book, *Turning to One Another*, bearing ORD's seal of approval on its cover.
Wheatley urged environmentalists to abandon Western science in favor of
New Science.[6] She taught that this experience initially leaves scientists in
a state of confusion, which she calls the "space of not knowing" and the
"abyss." While passing through the abyss, scientists shed their religious
beliefs and sexual inhibitions, emerging on the other side closer to nature
and one another.

Basing the future of EPA's research on some kind of new age environmental science that requires abandoning traditional science and religion
seems too far-fetched to be true. Apparently it's not. Longest proceeded
to contract the Naval Postgraduate School in Monterey, California, to
administer a questionnaire to ORD employees interested in moving up the
ladder.[7] Employees were informed that participating in his Appreciative
Inquiry Research Study, which included answering questions about their
religious beliefs and sexual desires and then being interviewed, involved
"no risks or discomforts." Anonymity was assured in that that their names
would be replaced with a confidential code. But they were compelled to

EXCERPT FROM THE ORD APPRECIATIVE INQUIRY
RESEARCH STUDY:

1	2	3	4	5	6	7

Strongly Disagree   Neither agree nor disagree   Strongly agree

38. I can think of several personality traits that turn me on sexually.

39. Taking care of others gives me a warm feeling inside.

40. Many people respect me.

41. I am full of gratitude.

42. The people around me make a lot of jokes.

43. I feel wonder almost every day.

44. I often feel curiosity.

45. My life is very fulfilling.

46. On a typical day, many events make me happy.

47. I often feel hopeful.

48. People are usually considerate of my needs and feelings.

49. I've often imagined being sexual with a friend, colleague, or acquaintance.

50. I often notice people who need help.

51. I always stand up for what I believe.

52. I am always respectful of people of higher status than myself.

53. I make jokes about everything.

54. I often look for patterns in the objects around me.

55. I learn something new every day.

56. I slow down to enjoy the moment whenever I can.

57. Good things happen to me all of the time.

58. I am not a quitter.

59. I love many people.

60. Almost everybody has something sexy about them.

sign a "Minimal Risk Consent Statement," and told that unforeseen circumstances may require decoding their names to "enhance the value of the
research data."

The whole exercise came to an abrupt end when some unappreciative
employee leaked the questionnaire to the *Washington Post*.[8] Interspersed
among seemingly innocuous questions, the questionnaire asked participants
to agree or disagree with statements of a sexual nature, such as, "I am a very
flirtatious person"; "When I'm attracted to someone, I am overwhelmed by
desire"; "Almost everybody has something sexy about them"; "I have dated a
lot of people"; "I have difficulty talking to attractive persons of the opposite
sex"; and "I become self-conscious when using public toilets." A number
of questions and statements also probed employees' religious beliefs, such
as, "Do you belong to a church, temple or other religious group?;" "I think
about God;" "I regularly attend religious services;" "I put my trust in God;"
"A higher power is looking out for me;" and "I seek God's help. Others tested
their loyalty, for example, "When someone criticizes ORD, it feels like a
personal insult;" "If a story in the media criticized ORD, I would feel embarrassed;" and "I am always respectful of people of higher status than myself."

Henry Longest's infamous Appreciative Inquiry appears to have been
an attempt to integrate Margaret Wheatley's views on environmental
science, sex, and religion when recruiting upper-level managers at ORD.
If so, it is an extreme example of how even scientists working for the US
government can still experience discrimination based on their religious
beliefs as well as research results. Unfortunately, Longest's interests in using
Wheatley's philosophies to build ORD's management structure conflicted
with my research documenting public health problems associated with the
agricultural use of processed sewage sludge. In her book *Turning to One
Another*, Wheatley stressed the importance of using wastes to produce food.
So, believe it or not, in the bizarre world of ORD management in the late
1990s to mid 2000s, the future of EPA scientists turned on their positions
regarding sewage, sex, and religion.

Religious Upbringing

My father, who was a Navy pilot, had no use for religion. He said he just
saw too much hypocrisy among chaplains during World War II. In the late
1930s, he trained with the Canadian Royal Air Force. Air Marshall William
Avery Bishop pinned his wings on just before he transferred to the RAF

in London. After Pearl Harbor, he signed up with the Navy and started training to fly PBY sea planes at the Naval Air Station in Pensacola, Florida. He eventually became a flight instructor as an all-weather-flying expert, and spent most of his time during the war flying high-ranking officers around when they needed to get somewhere regardless of the weather conditions.

He once delivered a dispatch to General Eisenhower's office, and stood just a few feet away from him. So, in memory of my father, I have included a portion of President Eisenhower's Farewell Address that directly relates to the subject of this book. The threat federal spending poses to academic freedom is "ever present" and "gravely to be regarded," Eisenhower said.

My mother saw to it that I attended Sunday School at the First Baptist Church of Thomasville while I was growing up in South Georgia in the early 1960s. In 1975, I was licensed as a minister in the second largest Southern Baptist church in Alabama. I never wanted to preach for a living, and, after about ten years of volunteering in churches, I could understand why my father felt the way he did. The same year I was licensed, the church in Alabama voted on whether to allow African Americans to join the church. Now, as I look back, that seems like a fitting beginning to many years of introspection. I eventually realized that there were too many inconsistencies between what I experienced in church, and what I believe is inherently right or wrong.

Lt. and Mrs. H. W. Lewis
Circa 1950–1953

Most of all, I couldn't reconcile what was preached from the pulpit versus what was practiced by the institution. Eventually the chasm between the church's views and mine became too wide to bridge. Nevertheless, it was insightful to observe how institutions dealing with religion and science both embrace different standards of ethics, one for individuals acting on their own, and another for those who serve the interests of the institutions. There's little difference between the government's manipulating research and targeting scientists who publish unwanted results versus churches and synagogues not practicing what they preach. Both are void of an essential ingredient of what I consider true science and religion, which is plain, old-fashioned ethics.

Ethics in Science and Religion

Based on religious beliefs, what boundaries should the government place upon scientific research? None, I would say, except that scientists should treat others as they want to be treated, which is the essence of ethics as well as religion. Scientists, no doubt, will continue to test the boundaries of knowledge in their quest to understand how the universe works. Inevitably, they will come into conflict with those who govern, whether by religious traditions, military force, or the will of the people. Still, so long as the goal is following the ethic of reciprocity, or Golden Rule, common sense and decency can prevail even when the most sacred beliefs and traditions are challenged.

Stem-cell research, for example, raises questions about what life is, and when it begins. Cells growing in a Petri dish, however, only have the potential to become people. Even those who interpret the Biblical story of Adam and Eve literally must agree that even a lump of clay has the potential to become a person. Of course, stem cells replicating in a culture have life, but so do viruses, which can be purified and stored in a bottle like crystals of salt until the sun consumes the earth. From the standpoint of the Golden Rule, a person comes into being at some point during prenatal development and ceases to exist at some point during the process of death. Because knowing how we would want to be treated requires a functioning brain, our existence as a person from the standpoint of the Golden Rule inextricably rests upon brain function.

Physicians who don't want to be kept alive past the point at which they can no longer function as a person should have no remorse over treating their patients likewise, if that's what their patients also want. The law, in other words, should not prevent people from giving up their lives so that those they love can go on with theirs. On the other hand, physicians who disagree should not be compelled to assist. The government should not stand in the way in either case unless, for some reason, it objects to people following the Golden Rule. Similarly, I expect that scientific studies could demonstrate that people with normal brains and the public interest at heart would want their own bodies to have been used to save others had they been aborted as fetuses. How is that any different than anyone, including Jesus Christ, giving his or her life to save others? It's simply treating others as they would want to be treated, is it not?

Some, of course, may argue that the Golden Rule cannot always be applied. For example, should a suicidal murderer seeking a showdown with police, i.e., "death-by-cop," be permitted to kill others because that's how he wants to be treated? Of course not. Although it's not always the case, the law of the land should be established by people of sound mind who have the public interest at heart. Most people meeting these requirements would want to be imprisoned, or even killed if necessary, if they ever became so morally corrupt, or insane, as to start shooting innocent people in schools and movie theaters.

With regard to research misconduct, the same standards applied to a single scientific article should be applied to the body of science as a whole. When authors are found to have selectively published data supporting their conclusions while disregarding unsupportive data, the whole paper is retracted, not just the portion containing fake data. Yet government agencies, corporations, and the academic institutions they fund appear to have no qualms about selectively funding scientists who support their interests, while taking out scientists who threaten them. The amount of resources that government agencies, universities, and corporations expend in making sure that the body of scientific literature as a whole supports government policies and industry practices is vast beyond imagination.[9]

Their objectives are to guarantee that most scientific studies agree with their positions, and that there is no documented evidence of any alleged problems with any of their products, policies, or practices. Achieving this position provides a powerful defense against lawsuits filed by plaintiffs seeking damages, and it wards off overly restrictive government regulations. Unfortunately, the body of scientific literature it produces is only an illusion.[10]

What if I held in one hand all of the peer-reviewed scientific articles in areas in which government and industry have heavily funded studies to support certain policies and practices, and suppressed the publication of studies yielding unfavorable results—and, in the other hand, held all of the articles retracted for selectively publishing only results that supported the authors' conclusions? Which has more credibility?

If we apply the same standard of ethics to both, there is no difference between the two. What this means is that we have no credible body of science in any area in which government and industry have a major stake in what gets published in the scientific literature. Another way to look at it is that the only areas in which we have a credible body of science are those that have no significant impact on government or industry, that is areas that don't really make much difference anyway in our daily lives.

To me, pure science has much in common with pure religion, at least as the book ascribed to James, the brother of Jesus, defined it.[11] *Pure religion* is simply helping the poor, oppressed, and needy, and not becoming entangled in "the world." The world, in his day, was progressing toward war between the Roman Empire and Israel's theocracy, neither of which cared very much about the poor and needy. If morality is measured in terms of caring for those who are least able to care for themselves, then, in the same vein, science should not be corrupted by government and industry to the detriment of public welfare.

Hence, just as pure religion is moral and untangled in the world, pure science is ethical and free from worldly corruption. In short, we are talking about the same standard of right and wrong for both science and religion. I think most people would agree that our religious and scientific institutions, for all the same reasons, have both fallen far short of meeting such a standard.

Another shortcoming of both science and religion is epitomized in the old adage "not seeing the forest for the trees." Most scientists tend to focus on one leaf or another, content to confine their work to some comfortable nook or cranny in science. But no one can really understand much about a leaf without comprehending how it functions as part of the whole tree, the forest, and even other forests around the planet. Many theologians, in my opinion, are even worse.

I often think back to something Clifford Baldowski, a noted political cartoonist for the *Atlanta Constitution*, once said. One of his sons was a college buddy of mine, and I spent countless joyful weekends with his family in my early years at UGA. One evening over dinner, "Baldy," as he was known, responded to his wife's supporting her point of view with a quote from scripture. He said something like, "Honey, the Bible can be used to support anything." For many years, I wondered why that is so characteristic of a book many believe to be more reliable than science. I eventually concluded it's simply because most religious leaders have never formed a big picture in which all of the parts of the Bible can fit without contradiction.

Absent any kind of overarching concept that encompasses all of its contents, any outlying statements are free to have any number of contradictory meanings. Generally, people just pick and choose whatever parts suit them, and just ignore the rest. I was never content with this approach, and spent several decades talking about it with David Gattie, a UGA professor of engineering who once worked with me at EPA.

Ancient religious texts, for example, the Tanakh (Torah, Prophets, and Writings in the Hebrew Old Testament), the Christian New Testament, and the Quran, require that their followers make certain moral choices: to judge or judge not; to love or hate. Influenced by their upbringing, life experiences, and whatever else motivates them, people use the ancient writings to establish their moral standards and religious beliefs.

In the Torah, for example, Moses, is credited with setting before the people a body of statutes that embodied both good and evil, which, correspondingly, would yield either life or death. They were free to decide whether to stone adulterers to death, or show mercy. Unfortunately, they chose the parts that led to death and, as a result, died in the wilderness. Likewise, Muslims must choose between following the verse in the Quran that praises Muslims for loving those who do not love them; or, they can kill unbelievers wherever they are found.[12] The instructions that people pick and choose to follow from whatever body of documents they consider to be sacred simply reflect whatever is in their hearts from the beginning. In other words, they use religious texts to morally justify whatever they have done, or want to do. To quote the author of the Book of Hebrews, the scriptures are "a discerner of the thoughts and intents of the heart."[13]

This is probably true of any collections of instructions that would cause someone to choose between what they consider to be right and wrong. For example, I could write "Stomp the ant" on one side of a piece of paper, on the other side write "Let the ant go," and hand it to a child watching an ant go by. Most children would probably do whatever they are already inclined to do, which, at least for most boys, is stomp the ant.

Although many religious leaders are known for cherry-picking the Bible and other religious texts to support their beliefs, the Tanakh, Christian New

Testament, and Quran all prohibit the practice, at least so far as the body of law is concerned.[14] According to the Hebrew Law of Moses, anyone who transgresses any part of it is guilty of transgressing all of it.[15] Hence, those who judge others as being guilty of idolatry, sexual perversion, murder, and other transgressions of the law without mercy, which the law requires, is condemned under the same law as an idolater, fornicator, and murderer himself.[16] In other words, because the law was written such that one can only keep all of it or none of it, anyone who judges someone without mercy is guilty of transgressing the whole law. Although he may have never murdered anyone, blasphemed God, or committed fornication, he is no different in the eyes of the law than those who have.

There are doubtless many reasons why religious texts contain conflicting instructions concerning morality and ethics. But two things stand out. As mentioned earlier, Moses gave those who followed him out of Egypt a choice of paths to follow, one good and the other evil. After delivering them out of slavery, he instructed them to worship only one God. Egypt experienced a similar uprising in its past when Pharaoh Amenhotep IV moved Egypt's capital into the wilderness, built the city of Armana, and instructed the people to worship only the creator Aten, the giver of life. His followers eventually ended up back where they started. Things didn't go well with Moses' followers, either. They later complained that they would have preferred that God kill them while they were eating in Egypt, instead of starving them to death in the wilderness.[17]

To deal with their rebellion, Moses re-created many of the trappings of Egypt's religious system, including priests, animal sacrifices, and an inner sanctum for the high priest to commune with the Lord of Israel. He also reinstituted a system of law familiar to the people of Egypt and its neighbors, which carried the penalty of death for blasphemy, murder, adultery, homosexuality, and other transgressions. But with these changes, Moses proclaimed that the Lord is full of mercy as well as vengeance.[18] This conflict between judgment and mercy, life and death, continues throughout the Tanakh, and carries over to the Christian New Testament and the Quran. James, for example, instructed the early Jewish converts that no mercy would be shown toward those who judge others without mercy.[19]

The struggle between judgment and mercy plays out in a multitude of contradictory commandments and teachings that are epitomized in the story of Jesus sparing a woman caught in adultery, saying, "He that has no sin, let him cast the first stone."[20] The purpose of the contradictions, as Moses indicated, was to force people to choose between good an evil, life and death—a choice that would come back to them in the end when God would judge them by whatever judgment they used to judge others.[21]

Religion versus Science

Another reason why religious texts contain conflicting instructions concerning morality and ethics is perhaps unintentional, and gives rise to many of the conflicts among religious groups, and between them and secular society. Mainstream religious leaders tend to believe that the Bible and other sacred texts deal with spiritual truths and are not meant to explain physical or biological phenomena. Fundamentalists disagree, and will cling to their beliefs no matter what science may demonstrate.

So far as science is concerned, most fundamentalists are comfortable living somewhere between the seventeenth and nineteenth centuries—after Galileo and before Darwin. They accept the fact that the sun doesn't revolve around the earth, but not that they evolved from some lowly creature that once crawled upon it. Their core beliefs haven't changed since the earliest times, when pagan priests attributed floods, droughts, and other natural disasters to an angry god punishing people for their transgressions. And, of course, the priests were never punished. Their gods always required the shedding of innocent blood instead.

The Bible as a whole reflects the evolution of religion from the dawn of civilization, when primitive cultures sacrificed humans and animals to appease their angry gods, to modern times, when people strive to rid the earth of slavery, genocide, torture, sexism, racism, murder, assault, fraud, deceit, and intimidation. Such was the life of the Apostle Paul. He stood with one foot in the ancient world of animal sacrifices, racial and sexual discrimination, and death by stoning for crimes that are not even recognized in modern society. The other was firmly planted in a new world in which laws and prophets were followed by simply treating others as we would want to be treated.[22]

When it comes to science, religious fundamentalists argue that most scientists aren't *believers*; therefore, they are children of the devil. Yet the Bible says, "You believe there is one God, you do well, [but] even the devils believe and tremble."[23] Likewise, Jesus said that many who call him Lord, and have worked miracles in his name, are workers of iniquity.[24] Fundamentalists believe that one day they will be like God, and reign in heaven. Isaiah said that Lucifer believed the same thing.[25] Fundamentalist Christians believe that Muslims and Jews are bound for hell. But the Bible says that one day the Lord will say, "Blessed be Egypt my people, and Assyria, the work of my hands, and Israel, my inheritance."[26]

Personally, I'm comfortable being like the man who brought his infirmed son to Jesus, saying, "I believe, help my unbelief;" or the Apostle Thomas, who could not believe in the resurrection without physical proof.[27] I'm comfortable calling Jews and Muslims and anyone else who loves his neighbor as himself my brother. And I am confident that Jesus, who said, "No man comes to the Father but by me," would judge them according to what in is their hearts, and receive them with open arms. To believe otherwise is to reject the very heart of what Jesus and his apostles taught, namely, that laws and the writings of prophets can be adhered to simply by loving others as ourselves.[28] As the Apostle John wrote, "Every one that loves is born of God, and knows God".[29] "Transgressing the law, or *sin*, therefore, is simply failing to treat others as we want to be treated."[30] "It is the law written not in stone, but upon our hearts".[31]

Synagogues, mosques, churches, and other religious institutions, therefore, have but one mission, which is to show the world that loving God means, *in all things*, treating others as we want to be treated, and caring for even the least wealthy and least powerful among us. To those theologians who disagree, I would point to what Jesus told Nicodemus: "You must be born again."[32] But not as evangelical Christians use the term. Instead, you must forget everything you think you know about right and wrong, and become like the little children who entered the Promised Land with no knowledge of good and evil.[33]

This, I believe, is the true meaning of the story of Adam and Eve in the Garden of Eden, in which humanity was separated from God by eating the fruit of the tree of knowledge of good and evil.[34] Everything written in religious writings that people use to judge one another comprises the knowledge of good and evil. In other words, the proverbial fruit Satan used to tempt Eve was the fruit of the world's religions. As James, the brother of Jesus, taught, to practice pure religion is to remain uncorrupted by the world, and simply treat even the least among us as we would have others treat us e.g., care for the fatherless and widows.[35] Redemption comes not from the works of our hands, but from the love we have in our hearts for one another.

Jesus told his disciples that he will separate the sheep from the goats one day, and take to himself those who fed the least of his brethren when they were hungry, gave them water when they were thirsty, took them in when they were strangers, clothed them when they were naked, and visited them when they were sick and in prison.[36] If that's true, I will feel good about having devoted my life to science. I cannot think of anyone who has fed more of the hungry, provided more water to the thirsty, sheltered more strangers, clothed more people who have nothing to wear, healed more of the sick, and freed more of the innocent from prison than scientists who have truly worked for the public good.

Although the two worlds of science and religion can never merge, I still think that scientists should consider that research ethics, which is rooted in religious beliefs about right and wrong, good and evil, is an inherent part of science. The ethical ramifications of cherry-picking scientific data is just one example in which scientists may benefit from a better understanding of the parallels that are found in religion.

Perhaps understanding some of these parallels would help those who run our scientific institutions appreciate the utter degradation of integrity that comes from sanctioning unethical behavior by those who act on behalf of their institutions, while punishing those who do the same to serve their own self-interests. And, by appreciating how the Law of Moses was designed such that it could not be violated in part without transgressing the whole, perhaps scientists will understand why allowing one scientist to be unjustly silenced can destroy the integrity of the whole body of science.

7

BRIAN DEER—HERO OR HOAX?

In December of 2012, Caroline Snyder, who founded Citizens for Sludge-Free Land, created a "We The People" White House petition to ban land application of biosolids.[1] Because many of the chemicals found in high concentrations in biosolids are linked to autism, I forwarded the petition to organizations involved with autism. In fact, the top ten groups of environmental pollutants linked to autism are all found concentrated in biosolids, which are applied to farms, parks, school playgrounds, and other public and private lands.

When bringing me up-to-date on events related to autism, several recipients of the petition commented that Brian Deer, a freelance reporter in the UK, had published allegations of research misconduct against me on his website. Deer's claim to fame comes from publishing similar allegations against Andrew Wakefield, who published a *Lancet* article associating autism with inflammatory bowel disease.[2] Though this association is now widely supported within the scientific community,[3] Wakefield and his coauthors came under fire because parents of eight of the twelve children in the study blamed their children's autism on the measles, mumps, and rubella (MMR) vaccine.

The UK General Medical Council (GMC) held hearings into Deer's allegations, which were first published by the *Sunday Times of London* in 2004.[4] It revoked the medical licenses of Wakefield and his senior coauthor, John Walker-Smith.[5] The High Court of England, however, eventually threw out the sanctions against Walker-Smith, describing the GMC panel's various findings as "not legitimate," "perverse," "unsustainable," and "untenable."[6] Wakefield had moved to the United States and dropped his appeal. He's currently suing Deer in the State of Texas.[7]

The British Government's Role

In 2004, Brian Deer and an influential member of Parliament, Evan Harris, met with editors of *The Lancet*, and accused Andrew Wakefield and his coauthors of research misconduct concerning the 1998 *Lancet* study. Harris wanted the Crown Prosecution Service to hold Wakefield and his coauthors responsible for the deaths of unvaccinated children and consider charging them under criminal statutes. When Parliament turned the matter over to GMC, Harris made sure that Wakefield and his coauthors would be held to a criminal standard, just in case he ever succeeded in gaining criminal prosecutions.

Behind the scenes, Parliament was dealing with a critical loss of public confidence in the MMR vaccine. From 1988 until 1992, SmithKline Beecham marketed the MMR vaccine, Pluserix, in Great Britain. It was withdrawn because the mumps component caused outbreaks of aseptic (viral) meningitis. Because meningitis involves brain inflammation, the MMR vaccine had already been implicated in the United States as a possible factor in causing autism.

At a press conference when the *Lancet* article was published, Wakefield was asked about the MMR vaccine controversy. He recommended that concerned parents talk with their pediatricians about having their children vaccinated with the single measles vaccine, which was available in Great Britain at the time. Reacting to parents, shifting to the single measles vaccine, the British Government withdrew it. With one MMR vaccine withdrawn, parents in the *Lancet* study blaming all commercially available MMR vaccines for causing autism, and the single measles vaccine without the mumps component no longer available, many parents chose not to vaccinate their children against measles.

While many parents blame the British government, the medical community holds Wakefield responsible for any subsequent measles outbreaks attributed to low vaccination rates. Others note that mumps, which is a mild disease in children, can cause serious complications in adults. Because mumps vaccines don't provide lasting immunity, older people are placed at increased risk of mumps infections.

This backdrop yields important context when considering the sequence of events that transpired with Brian Deer's involvement in the MMR controversy. It begins with Harris's accompanying him to lodge allegations of research misconduct with *Lancet* editors. These allegations, in turn, prompted Parliament to hold an inquiry and turn the matter over to the GMC.

Then we see GMC's prosecutors arranging to have their expert pediatric gastroenterologist, Professor Ian Booth, conduct a highly questionable analysis of the *Lancet* children's routine pathology reports, which made it appear that Wakefield exaggerated information in the grading sheets provided by one of his senior coauthors, Dr. Paul Dhillon, who was a world-renowned expert in examining colon biopsies.

Finally, we see the *British Medical Journal* (*BMJ*), which is sponsored by manufacturers of the MMR vaccine, publishing Booth's analysis just as the GMC hearings end. While the *BMJ* claims that the source of this analysis is an award-wining investigative reporter, Brian Deer, I discovered that the actual source was the GMC's solicitors acting in the interest of the British government. This raises an important question: Has Brian Deer simply pieced together bits and pieces of records from different sources to create a hoax? If so, it may be the greatest hoax perpetrated upon the scientific community since Charles Dawson created the infamous Piltdown Man.

Deer's Allegations against Me

As it turned out, the feedback I was getting from autism groups had to do with Deer's republishing the industry white paper that Synagro, a biosolids company, published in 2001.[8] Synagro filed its allegations as a research misconduct petition at the UGA, hoping it would force me to withdraw research papers that implicated the company's products in causing public health problems. The US EPA found that the allegations were not based in any facts, and the company withdrew its petition in 2004.[9] When Deer republished the company's allegations, he neglected to mention that I had been cleared of all the allegations.

Deer is miffed at me for publishing evidence used in the GMC's proceedings, which Dr. Wakefield provided from his court files. These documents, which were previously unpublished, raised serious questions about Deer's allegations of research fraud against Wakefield. As such, Dr. Wakefield was free to release them in his own defense under US law. Their authenticity was confirmed to me and others by various sources, including one of Wakefield's attorneys in the UK, the GMC's expert witness who created one of the documents, and one of Wakefield's coauthors;

they prompted the *BMJ* to revise its fraud allegations.[10] On his website, Deer used Synagro's white paper to allege that EPA fired me for research misconduct.[11] He also claimed that the Labor Department ruled against me in all of my whistleblower complaints. No one else has ever published such allegations, all of which are easily disproved by reading any of countless news articles, court transcripts, and congressional records readily available on the Internet. Deer also argued that I'm not even a legitimate whistleblower, and that the National Whistleblowers Center where I serve on the Board of Directors is just a front for a private law firm. These are only a few of Deer's allegations.

Deer's investigative reports about Wakefield and his coauthors also involve a plethora of false allegations. Oddly enough, legal experts say this approach offers some protection against libel lawsuits. When someone has a history of making false allegations, it's difficult to prove in a court of law that he or she is aware in any particular instance that he or she is, in fact, lying. Moreover, when someone is accused of numerous acts of wrongdoing, people tend to believe that there must be some truth to at least some of them.

To draw public attention to his allegations against me, Deer posted links to them in articles and comments he published in *Nature* and other scientific journals.[12] Fiona Godlee, the *BMJ*'s editor-in-chief, commended Deer for a job well done, and linked the *BMJ* to the section of Deer's website containing the allegations.[13]

Ed Hallman, the attorney who handled my qui tam cases, urged Godlee to remove the *BMJ*'s link to Deer's attacks:[14]

HALLMAN TO GODLEE (2013)

As Mr. Deer knows from reading the court records, EPA's retaliatory actions that led to Dr. Lewis' involuntary retirement had to do with a 2-page commentary Dr. Lewis published in Nature in 1996, entitled: "EPA science: Casualty of election politics." This commentary predates Dr. Lewis' research on sewage sludge. EPA retaliated over the commentary; and the Labor Department twice ruled in favor of Dr. Lewis. It was these Labor Department rulings in favor of Dr. Lewis that prompted two hearings by the US House of Representatives in 2000.

When falsely alleging that Dr. Lewis' termination had something to do with Synagro's allegations of research misconduct, Deer omits the inconvenient facts that the date of Dr. Lewis' departure from EPA was set by a settlement agreement that predates Synagro's allegations by three years. That fact alone condemns the truthfulness of Mr. Deer's allegations about Dr. Lewis, and BMJ's adoption of such lies....

Since Synagro and other industry websites removed the white paper, it is a mystery as to how Mr. Deer even obtained a copy of it. On his website, Deer simply re-publishes Synagro's original PDF files. They are not linked to any other source that may have posted them.

In what looks like an attempt to cover up the evidence if I were to file suit, Deer apparently inserted a typographical error into the links to Synagro's white paper published on his website. In April 2013, he personally emailed James Carter, one of the attorneys who communicated with him and Godlee on my behalf:[15]

DEER TO CARTER (2013)

(a) The link(s) to which you refer aren't operative, and, if they were ever operative (and I don't think they were), they didn't link to anything that wasn't posted many years ago, and hence isn't actionable under the relevant limitations statute.... However, as I say, the links are dead. I have records showing that they were dead a long time ago, if not always. If you look at the source code carefully, you will see that the "f" has been omitted from "pdf". Thus, the links wouldn't have worked.

About 356 results (0.23 seconds)

[PDF] Synagro White Paper Exec Summ _Working - Brian Deer
briandeer.com/solved/**david-lewis-synagro**-summary.pdf
File Format: PDF/Adobe Acrobat - Quick View
Analysis of **David Lewis'** Theories Regarding Biosolids ... **Synagro** presents evidence
in this **paper** of the weakness of Dr. Lewis' position to better enable ...

Google Search for Synagro White Paper. Courtesy of David Lewis.

My attorneys and I have been downloading files Deer has posted on his website since 2011, and converted his web pages to pdf files to preserve the internal links. There are multiple ways to prove that the typographical error to which Deer refers was introduced to his website shortly before he emailed Mr. Carter. For example, the Google search displayed above, which was downloaded in January 2013, shows that Deer's website was the number-one, and *only*, Google source of Synagro's white paper. In this record, the "pdf" file format at the end of Deer's URL is visible.

Deer offered to stop his attacks if I would refrain from making any further comments about has allegations of research fraud against Andrew Wakefield. He warned that, if I refuse his offer, he will release far more damaging information about me and the National Whistleblowers Center:[16]

> *DEER TO CARTER (2013)*
>
> *A few weeks back, I downloaded a large body of documents from Synagro's website, where they were posted for all to see.... I now have a substantial file on Dr. Lewis, including his various statements and depositions, along with what appear to be the key documents of the Marshall case. I must say that they reflect poorly on your client, as do the interminable court judgments against him. I believe his activities have the potential to cause great damage to the National Whistleblowers Center.... I propose to make no further comments about your client. If, however, he continues his campaign of public defamation against me, then a different situation arises....*

Trouble in Paradise

It all started back in January 2011, when Stephen Kohn, executive director of the National Whistleblowers Center, spoke at a vaccine safety conference in Jamaica.[17] At the last minute, one of the conference organizers offered to cover my plane ticket if I would talk about my experiences at EPA and comment on the technical presentations. A local resident let me use a spare bedroom. As a research microbiologist, I had a general knowledge of vaccines, but wasn't aware of any issues surrounding vaccine safety.

During the time of the conference, *BMJ*'s editors released a series of articles by Brian Deer accusing Andrew Wakefield of research fraud. CNN's Anderson Cooper interviewed Wakefield by phone while Wakefield attended the conference. Cooper followed up with interviews of Sanjay Gupta, Bill Gates, and Brian Deer.[18] With a leading medical journal backing Deer's allegations, and the GMC having sanctioned Wakefield for ethics violations, global news organizations and virtually every major science and medical journal published the allegations without questioning their veracity.

Before leaving the conference, Wakefield offered to give everyone a free copy of his book, *Callous Disregard*, which addressed Deer's earlier allegations published in the *Sunday Times*. I introduced myself, and asked him to mail me a copy. After reading *Callous Disregard* and the various articles published by the *BMJ*, I contacted Wakefield to see whether he still had any of the original documents from the GMC's proceedings.

I was particularly interested in any documents related to the diagnosis of inflammatory bowel disease (IBD) published in Table 1 of his *Lancet* article. In 2010, the *BMJ* published an article by Deer titled "Wakefield's 'Autistic Enterocolitis' Under the Microscope." In it, Deer accused Wakefield of exaggerating the diagnoses of nonspecific colitis, a form of IBD, in the *Lancet* children. Wakefield summarized the diagnoses in Table 1 of the *Lancet* article.[19] The grading sheets, which Wakefield used to create the table, were completed by two coauthors, professor A. P. Dhillon and Dr. Andrew Anthony.

Pathology Grading Sheets

In his "Autistic Enterocolitis" article, Deer wrote that pathology grading sheets "don't generate clinical diagnoses such as colitis."[20] Hence, for Wakefield to use the grading sheets to conclude that the *Lancet* children had colitis, he must have made up the diagnosis. According to Deer, the only way anyone could prove Wakefield did, in fact, fabricate the diagnosis was to reexamine the children's biopsy slides, which contained tissue samples collected from the children's colons using flexible endoscopes. They are the "ultimate proof," Deer wrote, but, unfortunately, the slides are "missing."[21]

According to Deer, the routine pathology reports indicated that only three children had evidence of colitis. Table 1, which Wakefield created, indicates that eleven of the twelve *Lancet* children had colitis.

Deer wrote:

> *And, if Dhillon is right in saying the slides can't be found, the ultimate proof is missing. All we have are the pathology reports, which independent specialists seem to agree are largely unremarkable. "They wanted this bad," commented Tom MacDonald, dean of research at Barts and the London School of Medicine and coauthor of Immunology and Diseases of the Gut. "If I was the referee and the routine pathologists reported that 8/11 were within normal limits, or had trivial changes, but this was then revised by other people to 11/12 having non-specific colitis, then I would just tell the editor to reject the paper.*

In studies I've published on infection control with flexible endoscopes, biopsies were collected and examined for inflammation in over half of the procedures.[22] So, I'm familiar with the procedures involved in this part of the *Lancet* study. In my opinion, the biopsy slides would not be very helpful in resolving the issue of research fraud. Different pathologists examining the same biopsies often disagree over various degrees of mild to moderate inflammation, and their clinical significance.

To get around this problem, professor John Walker-Smith, a senior author of the *Lancet* study, brought in Dr. Amar Paul Dhillon, an internationally recognized expert in colon pathology, to conduct a blinded independent analysis of the children's biopsies using a standardized protocol. Dhillon was assisted by a junior pathologist, Dr. Andrew Anthony. The only other pathology data available were routine reports from the Royal Free Hospital. Those reports only reflected the non-expert opinions of on-duty pathologists who happened to be working on different shifts when the biopsies were collected. Based on Dhillon's blinded analysis, Walker-Smith disagreed with some of the diagnoses reported by the on-duty pathologists. Although Deer later accused Wakefield of making the changes as part of an "elaborate fraud," Walker-Smith testified at the GMC hearings, "This was totally unrelated to Andy Wakefield.... [It] was a personal initiative which I took myself...."[23] Dhillon responded to the *BMJ* publishing some of his grading sheets, which I obtained from Dr. Wakefield's files and submitted to the journal in 2011. He wrote, "It is not unusual for the clinical significance of microscopic observations to be reinterpreted and altered by this process, and it could be that the histological diagnostic interpretation subsequently has to be corrected."[24]

Some on-duty pathologists rated most of the children's biopsies as "normal." Walker-Smith, however, testified that the histology of the children's biopsies "did show evidence of both chronic inflammation and episodes of acute inflammation as instanced by cryptitis etc."[25] Similarly, in his response to the *BMJ* in 2011, Dhillon cautioned against using isolated documents to rule out colitis, and stated "In the context of a comprehensive clinico-pathological review by trusted clinical colleagues, the designated diagnosis of colitis seemed to me to be plausible."[26] In light of all the statements made by Dr. Wakefield's coauthors, both at the GMC hearings and, more recently, to the *BMJ*, one thing is clear. Allegations by Deer and the *BMJ*'s editors that Wakefield simply fabricated the diagnoses of colitis reported in the *Lancet* article are untrue.

Because Deer didn't indicate that he ever had anyone examine Dhillon's and Anthony's grading sheets, I assumed that they were also missing. I was surprised, therefore, when Wakefield offered to scan the GMC's copies of the grading sheets and email them to me. I was even more surprised when he said he could also send me the actual photomicrographs of the biopsy slides for at least six of the twelve children (Patients 2–6 and 9). Wakefield kept these photographs taken by Dhillon and Anthony to include in his lectures.

So far as evaluating the fraud allegations go, Dhillon's and Anthony's photomicrographs are much more useful than the biopsy slides. That's because they show the specific areas that Dhillon and Anthony considered indicative of inflammation. As such, they can be directly compared with the grading sheets to verify whether the cell structures described in the grading sheets are visible in the photomicrographs. If they aren't, the accuracy of the grading sheets is in doubt. But if they are, the accuracy of Table 1 of the *Lancet* article can be assessed by simply comparing Table 1 with the grading sheets.

Although I'm not qualified to assess the various levels of inflammation, I can look for the specific features Dhillon and Anthony described in the grading sheets to see whether they are visible in the corresponding photomicrographs. After comparing the two, I was satisfied that the grading sheets do, in fact, match what Dhillon and Anthony saw in the biopsy slides.

Allegations of Elaborate Fraud

Deer's article, "Wakefield's 'Autistic Enterocolitis' Under the Microscope," was published in April 2010, six weeks before the GMC sanctioned Wakefield and Walker-Smith. As soon as the sanctions were announced, *Lancet*'s editors quickly retracted their study, which was still generating public concerns over the MMR vaccine. Possibly timed to ride the media wave about to be created by the GMC, the *BMJ* opened the door for Deer to pick up the ball and run with newly minted allegations of research fraud.

Unfortunately, Deer's article was a complete flop at the box office. It was soon forgotten. In January 2011, Godlee gave Deer a second chance while Wakefield was speaking in Jamaica. This time the *BMJ* delivered a bundle of Deer's short stories, diapered with editorials and commentaries. It all grew from the tiny seed of fraud, which had required a microscope to dissect nine months earlier. So, with CNN and other media organizations lined up to cover the big event, Godlee gave birth to an elaborate fraud, screaming for attention.

Godlee described the new arrival's elaborate features when it was brought back later that year to graft on twelve additional appendages—Wakefield's coauthors:[27]

> *Our coverage in January showed how Wakefield manufactured the appearance of a link between the vaccine and regressive autism while employed by lawyers trying to build a case against the MMR vaccine, and while negotiating extraordinary commercial schemes that would succeed only if confidence in the vaccine was damaged. The articles, by investigative journalist Brian Deer, also showed that the conflicts of interest were not confined to Wakefield. They drew in his then employer, the Royal Free hospital and medical school. Now part of University College London, the Royal Free issued public statements of support for national immunisation policy while privately holding business meetings with Wakefield over purported diagnostic kits, single vaccines, and autism products meant to be sold on the back of the vaccine crisis.*

Fiona Godlee and her coeditors explained what drove their renewed assault on Wakefield:[28]

The Lancet *paper has of course been retracted, but for far narrower misconduct than is now apparent. The retraction statement cites the GMC's findings that the patients were not consecutively referred and the study did not have ethical approval, leaving the door open for those who want to continue to believe that the science, flawed though it always was, still stands. We hope that declaring the paper a fraud will close that door for good.*

Professor Booth's Expert Report

In 2011, Godlee and her coeditors praised Deer, who has no training in science or medicine, for coming up with the idea of comparing the children's medical records with the *Lancet* article:

> *The GMC launched its own proceedings that focused on whether the research was ethical. But while the disciplinary panel was examining the children's medical records in public, Deer compared them with what was published in the* Lancet. *His focus was now on whether the research was true.*

But, as I searched through copies of the GMC's materials contained in Wakefield's personal files, I came across a document that left me stunned. It was a confidential report by Professor Ian Booth, the GMC's expert in pediatric gastroenterology, which compared the Royal Free Hospital's routine pathology reports with Table 1 in the *Lancet* article, which Wakefield created from Dhillon's and Anthony's grading sheets.[29] His report, which was prepared in 2006, was a perfect match for the "Autistic Enterocolitis" article Deer published four years later in the *BMJ*.

In 2011, I posted a copy of Booth's report on my website at the National Whistleblowers Center (www.researchmisconduct.org). Deer immediately emailed the director, Stephen Kohn, stating that the *BMJ*'s finding of research fraud against Wakefield was not "critically dependent on mismatches in histopathology records" and requested "Dr Lewis to take down his pages."[30] He claimed that he was unaware of Booth's report until I posted it, but would not say how he first came up with the idea of using the routine pathology reports to accuse Wakefield of exaggerating the diagnoses in Table 1 of the *Lancet* article.

Upon comparing the routine pathology reports with the *Lancet* article, Booth reported that only three of eleven children had abnormal biopsies. Because Wakefield reported in the *Lancet* article that only one of the twelve children had a normal histology, Booth concluded the *Lancet* data were "altered" to present "an exaggerated view of the histology." "Scientific fraud," he wrote, could not be ruled out.[31]

Just to be clear, here's what happened, step by step:

1. The goal was to determine whether, in Table 1 of the *Lancet* article, Wakefield exaggerated the number of children who the expert's grading sheets indicated were suffering from colitis.
2. Booth, and then later Deer, decided *not* to use the expert's grading sheets, which Wakefield used to create Table 1. The grading sheets, according to Wakefield, indicated that only one child was completely normal, and Dhillon and Anthony approved Wakefield's conclusions.[220]
3. Instead, they used routine pathology reports prepared by non-experts with no blinded analysis, and no standardized protocol—which weren't used to create Table 1. According to them, these reports indicated that eight children were normal.
4. Because the routine pathology reports didn't match Table 1, they concluded that Wakefield exaggerated the data in the expert's grading sheets.

To underscore my point, let me illustrate it visually. Table 1 on the next page represents Table 1 of the *Lancet* article, which was created by Wakefield and approved by Dhillon and Anthony. Table 2 represents the routine pathology reports created by doctors at the Royal Free Hospital who weren't necessarily experts in inflammatory bowel disease, didn't participate in the study, and didn't follow the study protocol. Deer and the *BMJ* claim that Table 2 demonstrates that Table 1 is *exaggerated*.

Data produced in a blinded study using a standardized protocol with control samples (e.g., Wakefield et al.'s Table 1) could be used to assess the accuracy of biopsy reports produced by routine pathologists working in a clinical setting (e.g., Deer's Table 2)—but never vice versa.

It's understandable why a reporter untrained in science and medicine may use routine pathology reports to judge the accuracy of a table created from an expert histopathologist's standardized grading sheets. But, why would the GMC's expert pediatric gastroenterologist, Professor Ian Booth, do this? To me, that was the biggest mystery.

TABLE 1. EXAGGERATED?		TABLE 2. ACCURATE?	
Patient No.	Diagnosis	Patient No.	Diagnosis
1.	Colitis	1.	Colitis
2.	Colitis	2.	Colitis
3.	Colitis	3.	Colitis
4.	Colitis	4.	Normal
5.	Colitis	5.	Normal
6.	Colitis	6.	Normal
7.	Colitis	7.	Normal
8.	Colitis	8.	Normal
9.	Colitis	9.	Normal
10.	Colitis	10.	Normal
11.	Colitis	11.	Normal
12.	Normal	12.	Normal

It's reasonable to assume that had the grading sheets indicated fraud, the prosecutors would certainly have used them. So, I sent Professor Booth a copy of his expert report, and asked why he used the Royal Free Hospital's routine pathology reports instead of Dhillon's and Anthony's grading sheets. He explained that the GMC's prosecutors requested this analysis of the case records (i.e., routine pathology reports) and used it to support their case:[32]

BOOTH'S REPLY (2011)

> *Yes, this is my document, although my understanding is that its contents remain confidential between myself and the GMC's solicitors to whom I submitted it. My analysis of the case records of the children presented in the Lancet publication was carried out specifically at the request of the GMC's solicitors and it formed part of the basis of the case brought against Wakefield et al. by the legal team acting on behalf of the GMC.*

Wakefield informed me that Booth's report was never admitted as evidence. I searched the transcripts of the GMC hearings, and concluded that

Wakefield was right. I found no reference to Booth's analysis of the case reports (i.e., routine pathology reports) in which Booth concluded, "It is not therefore possible on the basis of the information I have seen to exclude scientific fraud in this area of the *Lancet* publication." As quoted above, the *BMJ* maintained that, when the GMC examined the routine pathology reports, it "focused on whether the research was ethical," whereas Deer "compared them with what was published in the *Lancet*"...and focused on "whether the research was true."[33]

Obviously, the *BMJ* was mistaken. The transcripts of the hearings show that the GMC did indeed analyze the pathology reports ad nauseam with respect to ethics, that is, whether authors of the *Lancet* article had obtained all of the proper ethics approvals to perform the kinds of tests reflected in the pathology reports. But its expert witness, Professor Booth, also compared them with Table 1 of the *Lancet* article to see whether research fraud could be excluded. Booth's email to me states that the GMC asked him to perform this analysis, that is, compare the case reports with the *Lancet* article to see whether they match the diagnoses of colitis listed in Table 1; he reported back to the GMC's solicitors that he could not exclude fraud. Deer, therefore, was not the only person who compared the routine pathology reports with the *Lancet* article in an attempt to determine whether the research was "true."

As discussed earlier, to see whether Dr. Wakefield exaggerated any records to create Table 1, they should have used the grading sheets that Wakefield used to create Table 1. Even then, it would be difficult at best to conclude whether any fraud occurred. The grading sheets document various types and numbers of certain cells that are considered indicative of colitis. Not surprisingly, opinions vary among different researchers as to the significance of various cells and numbers of cells with respect to colitis. Just because a doctor disagrees with a diagnosis in Table 1 doesn't prove that fraud occurred. It just means that two doctors don't agree on a diagnosis. That happens all the time. Even Ingvar Bjarnason, who wrote one of the commentaries Godlee used to comment on Dhillon's grading sheets, cautioned that they "don't clearly support charges that Wakefield deliberately misinterpreted the records. The data are subjective. It's different to say it's deliberate falsification."[34]

Because the GMC's prosecutors never introduced Booth's analysis during the hearings, they obviously decided that it wasn't a good idea to build a fraud case based on comparing either the grading sheets or the routine pathology sheets, as Booth and Deer did, with the *Lancet* article. The *BMJ*, as mentioned earlier, claimed that the GMC just focused on ethics and not fraud, whereas Deer focused on using the medical records to investigate fraud. Booth's expert report, and his email to me stating that the GMC's solicitors asked him to perform the analysis, show that the GMC considered the same evidence Deer and the *BMJ* would come to publish four years later, but never used it. In other words, the GMC had already considered the analysis used by Deer and the *BMJ* to accuse Wakefield of research fraud, and tossed it in the trash can, so to speak. How Deer and the *BMJ* came to pull it out of the *BMJ*'s trash and turn it into a second British Press Award for Brian Deer remains a mystery.[35]

In his expert report, Professor Booth only stated in his report that he couldn't *exclude* fraud. That's all anyone can say even when comparing the grading sheets with the diagnoses reported in the *Lancet* article. Comparing the diagnoses with what's written in routine pathology reports is even less conclusive so far as fraud is concerned, because Table 1 is based on the grading sheets and not routine pathology reports. So what can be concluded? Nothing with respect to scientific fraud, except that it can neither be ruled in nor out.

Deer wrote in his "Autistic Enterocolitis" article, "And, if Dhillon is right in saying the slides can't be found, the ultimate proof is missing. All we have are the pathology reports, which independent specialists seem to agree are largely unremarkable."[36] But, when confronted with Booth's report on my website, Deer appeared to contradict his statement that said, "All we have are the pathology reports." He wrote to Mr. Kohn, "Nor is the finding by the editors of the *BMJ* of research fraud against Wakefield critically dependent on mismatches in histopathology records, as alleged by Dr Lewis. Although Wakefield undoubtedly falsely reported the recorded gut pathology in the children...."[37] Similarly, when the *BMJ* published some of the documents I submitted, *Nature* reporter Eugenie Reich quoted Godlee: "The journal's conclusion of fraud was not based on the pathology but on a number of discrepancies between the children's records and the claims in the *Lancet* paper."[38] Even here, Godlee admits that the only evidence the *BMJ* had to accuse Wakefield of research fraud was the children's medical records. Again, the *BMJ*'s own expert, Ingvar Bjarnason, wrote that the diagnoses of non-specific colitis reported in the *Lancet* article "don't clearly support charges that Wakefield deliberately misinterpreted the records. The data are subjective."[39] Deer and the *BMJ*, therefore, had no basis for alleging that Dr. Wakefield committed research fraud.

The discovery of Booth's report, however, does raise the possibility of journalistic misconduct. Contrary to what the *BMJ* claimed, the GMC and

Deer both compared the same records as Booth for the same purpose. The only difference was the outcome. Booth found that research fraud *couldn't be excluded*. That's no more conclusive of research fraud than my comparing my license plate with the *Lancet* article and finding that I cannot exclude research fraud. By using the same kind of analysis Booth used, Deer and the *BMJ* are alleging that Wakefield "faked the link" between MMR and autism.

We are now left with some important questions concerning why anyone would even use a comparison of routine pathology reports with the children's diagnosis in Table 1 of the *Lancet* article to accuse Wakefield of research fraud.

First, here's what we've learned:

- According to Professor Booth, the GMC's solicitors asked him to perform the analysis. He wrote in his email to me, "My analysis of the case records of the children presented in the *Lancet* publication was carried out specifically at the request of the GMC's solicitors and it formed part of the basis of the case brought against Wakefield et al. by the legal team acting on behalf of the GMC."[40] He found that scientific fraud *could not be excluded.*
- According to Deer, he used this analysis because he had nothing else to use. He wrote in his "Autistic Enterocolitis" article, "And, if Dhillon is right in saying the slides can't be found, the ultimate proof is missing. *All we have are the pathology reports* [emphasis added], which independent specialists seem to agree are largely unremarkable."[41] Deer and the *BMJ* used the analysis to conclude that Wakefield was guilty of "faking the link" between MMR vaccines and autism.
- Finally, Booth's expert report and email to me demonstrate that the *BMJ* was wrong to portray Brian Deer as an ingenious reporter who came up with an analysis that never occurred to the GMC during the four years it held proceedings.

Now, here are some important questions that remain:

- How did Brian Deer, who is supposed to be a top investigative reporter, obtain the children's pathology reports and all of the other records he used to write his articles, but never saw Booth's expert report or any of the GMC's copies of the grading sheets?
- Deer claims that he obtained the children's medical records through discovery in a defamation lawsuit that Dr. Wakefield filed against him.

If that's true, then shouldn't he have all the same documents I obtained from Dr. Wakefield's files, which includes Booth's expert report and the GMC's copies of Dhillon's and Anthony's grading sheets?

To say the least, it's very difficult to understand how Deer investigated Wakefield's case for seven years (2004–2011) and yet never obtained such key documents that only took me a few weeks to locate after meeting Dr. Wakefield in 2011. The fact that these documents seriously call into question the allegations of research fraud published by Deer and the *BMJ* makes it look all the more suspicious. Consequently, Deer and the *BMJ* should provide some credible answers and supporting documentation to explain how they overlooked these documents for so long. They should also explain why they still accuse Dr. Wakefield of research fraud when the GMC's own expert correctly concluded only that fraud couldn't be excluded by comparing the children's medical records with the diagnoses of colitis reported in the *Lancet* article.

Even if Deer never saw Booth's report before I published it, someone supporting the GMC's case may still have suggested to Deer or the *BMJ* that they perform the analysis. After all, that's what the GMC's lawyers did with Booth. Why not with Deer or the *BMJ*? One thing I'm convinced never happened is that Deer came up with this idea all by himself. How long does a physician have to listen to a newspaper reporter talk about medicine before the physician can accurately gauge the reporter's level of knowledge and understanding? Or, how long does a professional auto mechanic have to listen to a reporter talk about repairing car engines, or a professional horse trader to listen to a reporter talk about trading horses, to know whether the other person knows what he or she is talking about? In my opinion, there's no way that Brian Deer, on his own, came up with the idea of performing the same analysis that the GMC's lawyers gave Booth to perform. Based on his interactions with my attorneys, and the presentations I've watched him give, he has little comprehension of scientific research.

Searching for Answers

In September 2011, I submitted confidential GMC documents to Fiona Godlee to consider publishing in the BMJ. They would end up forcing the *BMJ* to either bail out on Deer, or accuse Wakefield and his coauthors of an even more elaborate fraud. Alas, the strings that bound this tangled web grew even stronger.

DOCUMENTS SUBMITTED TO THE BMJ *(2011)*

1. Dhillon's and Anthony's grading sheets
2. Photomicrographs of the *Lancet* children's missing biopsy slides
3. Anthony's PowerPoint presentation concerning the biopsy slides
4. Dhillon's and Anthony's affidavits, stating that they fully approved the *Lancet* manuscript prior to submission
5. Professor Ian Booth's 2006 expert report, which matched Deer's analysis of routine pathology reports published by the *BMJ* in 2010 to 2011.
6. A three-thousand-word word commentary in which I discussed how these documents negated the *BMJ*'s allegations of research fraud

Dhillon's and Anthony's affidavits stated that they both reviewed and approved the patients' diagnoses that Wakefield summarized in Table 1 of the *Lancet* article. Anthony even wrote "colitis" in the margins of his grading sheets. And, in his PowerPoint presentation, Anthony methodically explained how the evidence in the photomicrographs clearly demonstrated colitis in the *Lancet* children's biopsies.

The documentary evidence I submitted to the *BMJ* put Godlee and the *BMJ* squarely between the proverbial rock and a hard place. They could publish all of it and distance themselves from Deer's fraud allegations, or, they could amend their fraud theory and claim that Wakefield's coauthors and others conspired to fake the diagnoses of colitis. But that's highly unlikely. Dhillon, Anthony, and ten other coauthors would have had to invent an incredible web of lies in their affidavits and oral testimony in the GMC's hearings.

Upon consulting with their lawyers, the *BMJ* chose the latter approach. In 2011, Godlee borrowed the term "institutional research misconduct" from my website to write an editorial accusing University College London and the authors of the 1998 *Lancet* article of conspiring to discredit MMR vaccines manufactured by Merck and GSK, which sponsor the *BMJ*, and market their own measles vaccine. The *BMJ* had no interest in publishing any of the other compelling evidence I submitted, which clearly demonstrated that its allegations of research fraud were false. This included Anthony's grading sheets, where he wrote "colitis" in the margins, the photomicrographs of the missing biopsy slides, and Booth's analysis, which perfectly matched Deer's award-winning "investigative" reporting—except that it came to a completely different conclusion.

After choosing to make Wakefield's elaborate fraud even more elaborate, deputy editor Anthony Delamothe emailed me: "We care about getting your opinion on the interpretation of the biopsies into the journal, but nothing more."[42]

The *BMJ*'s editors and lawyers then proceeded to draft a five-hundred-word "letter" for me to approve, which they published along with Godlee's editorial. They claimed it was peer-reviewed, but would not let me see the comments. I was only allowed to make minor changes in the letter, which had to be approved by the *BMJ*'s attorneys.

Draft copies of *my* letter stated that it would be published as my "rapid response" to Deer's "Autistic Enterocolitis" article, which is the way I submitted the rapid response I wrote. But when they published the letter they crafted to replace my rapid response, they published it as my response to an article by Deer titled "How the Case against MMR Vaccine Was Fixed."[43] That article contained none of Deer's allegations about Wakefield's misinterpreting the grading sheets, and how the missing biopsy slides—which I now had photomicrographs of—were the "ultimate proof" of Wakefield's guilt or innocence.

The other information Deer presented in "How the Case against MMR Vaccine Was Fixed" was based on the recollections of some of the parents of the *Lancet* children many years after the *Lancet* study was published. It was also based on patient records that were no more scientifically reliable for assessing the accuracy of diagnoses summarized in Table 1 of the *Lancet* article than the routine pathology reports and other general practice records that were unavailable to the authors of the *Lancet* study. These are the same records that the GMC debated over in an effort to determine whether Wakefield and two of his coauthors were guilty of ethics violations. Judge John Mitting of the High Court of England threw out the prosecution's arguments based on these records, stating:[44]

> As is apparent from the extracts summarized above, the medical records are equivocal. They do not point clearly either to the undertaking of a research project or to clinical diagnostic investigation. Both sides were able to make cogent submissions on the basis of the contents of the medical records to support their respective propositions.

Deer and Godlee now want to take these same records, which contain the physician's notes on what medical tests were done, when they were done, why they were done, and what the results were, and claim that they support their arguments that Wakefield intentionally misinterpreted Dhillon's and Anthony's grading sheets. That's absurd.

In November 2011, the *BMJ* published only Dhillon's grading sheets, along with the following articles discussing their significance:

1. **Editorial**: F. Godlee, "Institutional Research Misconduct."
2. **Feature article**: B. Deer, "Pathology Reports Solve 'New bowel Disease' Riddle."
3. **Commentary**: Ingvar Bjarnason, King's College Hospital, London, "We Came to an Overwhelming and Uniform Opinion That These Reports Do Not Show Colitis," Ingvar Bjarnason, King's College Hospital, London.
4. **Commentary**: Karel Geboes, Department of Pathology, KULeuven, Belgium, "I See No Convincing Evidence of 'Enterocolitis,' 'Colitis,' or a 'Unique Disease Process,' Karel Geboes, Department of Pathology, KULeuven, Belgium.

Godlee crystallized it all:

> *This week we publish new information that puts the spotlight on Wakefield's coauthors. Previously unpublished histopathology grading sheets apparently completed by Amar Dhillon, the senior pathologist on the paper, remove any remaining credibility from the claim that the Royal Free doctors had discovered a new inflammatory bowel disease associated with MMR. Along with UCL's failings during and after Wakefield's tenure, this evidence also raises wider concerns about the prevailing culture of Britain's academic institutions.*

After proclaiming that Dhillon's grading sheets prove that Wakefield's conspiracy was much wider than previously thought, Godlee used her editorials to publicly call upon Parliament and University College London to hold inquiries. UCL declined to investigate Wakefield for research misconduct, explaining, "The net result would likely be an incomplete set of evidence and an inconclusive process costing a substantial sum of public money."[45]

Keeping Them Honest

In closing, I would like to revisit Godlee's statements summarizing the implications of the grading sheets I submitted to the *BMJ*:

> *This week we publish new information that puts the spotlight on Wakefield's coauthors. Previously unpublished histopathology grading sheets apparently completed by Amar Dhillon, the senior pathologist on the paper,* **remove any remaining credibility from the claim that the Royal Free doctors had discovered a new inflammatory bowel disease associated with MMR** *[emphasis added].*

One can never overemphasize the importance of carefully reading the various related documents people draw upon to support their positions. So, let me close by pointing out that Godlee and Deer are conflating two completely different *Lancet* studies in order to support their fraud allegations.

1. In 1995, Wakefield coauthored a *Lancet* study reporting an increased prevalence of certain forms of IBD (Crohn's disease and ulcerative colitis) among individuals vaccinated with live measles vaccine compared with an unvaccinated group.[46] Other researchers found no association. A review published in 2001 stressed, "While further research is necessary into the causal factors underlying Crohn's disease and ulcerative colitis, continued public education efforts are needed to reassure the public about vaccine safety and to prevent declines in vaccine coverage."[47]
2. In the 1998 *Lancet* article, Wakefield and his coauthors reported a new IBD associated with Autism Spectrum Disorder (ASD), *not MMR*:

WAKEFIELD ET AL., THE LANCET *(1998)*

We describe a pattern of colitis [IBD] and ileal-lymphoidnodular hyperplasia in children with developmental disorders [ASD]. Intestinal and behavioural pathologies may have occurred together by chance, reflecting a selection bias in a self-referred group; however, the uniformity of the intestinal pathological changes and the fact that previous studies have found intestinal dysfunction in children with autistic spectrum disorders suggests that the connection is real and reflects a unique disease process.

3.   The association between IBD and ASD reported by Wakefield *et al.* is supported by a number of studies. See, for example, the two recent studies below, and their citations, which include some of Wakefield's papers:

   a.   Citing papers by Wakefield and his coauthors in 2011, for example, Williams and others at Columbia University and Harvard Medical School concluded that gastrointestinal disturbances are commonly reported in children with autism and may contribute to behavioral impairment.[48]

   b.   In a 2013 study titled "Identification of Unique Gene Expression Profile in Children with Regressive Autism Spectrum Disorder (ASD) and Ileocolitis," S. J. Walker at Wake Forest University and his coworkers independently concluded that the autism-related intestinal pathological changes reported by Wakefield and his coauthors are, in fact, real and do reflect a unique disease process.[49] With regard to ASD manifesting mucosal inflammatory infiltrates of the small and large intestine (ASDGI), the authors state: "ASDGI children have a gastrointestinal mucosal molecular profile that overlaps significantly with known inflammatory bowel disease (IBD), yet has distinctive features that further supports the presence of an ASD-associated IBD variant, or, alternatively, a prodromal phase of typical inflammatory bowel disease."

4.   In the 1998 *Lancet* article, Wakefield and his coauthors stated that there was *no scientific evidence* to support the parents' claims that autism in eight of the *Lancet* children was caused by MMR vaccination:

We did not prove an association between measles, mumps, and rubella vaccine and the syndrome described.... If there is a causal link between measles, mumps, and rubella vaccine and this syndrome, a rising incidence might be anticipated after the introduction of this vaccine in the UK in 1988. Published evidence is inadequate to show whether there is a change in incidence or a link with measles, mumps, and rubella vaccine.

Conclusion:

When using the grading sheets to support her fraud allegations, Godlee falsely stated that Wakefield and his coauthors concluded in their 1998 *Lancet* article that MMR vaccine is linked to autism. The media followed suit.[50] CNN's Anderson Cooper, for example, announced:

> *We begin, though, as always, "Keeping Them Honest." Breaking news tonight: Just hours ago, The British Medical Journal, (BMJ), did something extremely rare for a scientific journal. It accused a researcher, Andrew Wakefield, of outright fraud.*
>
> *Now, Wakefield is not just any researcher. His 1998 study on autism and childhood vaccines literally changed the way many parents think about vaccines. The study was based on just 12 children. That's right, 12 children. But many parents desperate for answers around the world embraced Wakefield's claim that he had found a link between autism and the vaccine for measles, mumps, and rubella.*

In the end, *BMJ* and the news media—aided by the GMC's lawyers—conflated two *Lancet* studies in order to disgrace the authors of the 1998 article for reporting the following two conclusions, both of which are true:

- Autism is associated with a unique form of inflammatory bowel disease.
- Some parents associate autism with MMR vaccination.

Does MMR Vaccine Cause Autism?

In 1998, Dr. Andrew Wakefield and his twelve coauthors at University College London reported in their article in *The Lancet* that parents of eight of the children in the study noticed a temporal association between MMR vaccination and autism spectrum disorder (ASD). Before that, numerous researchers reported that live rubella and measles viruses occasionally cause regressive ASD in young children. The validity of these earlier studies, so far as I can tell, has never been refuted. In 1991, the prestigious US Institute of Medicine (IOM), part of the National Academy of Sciences, put the medical community on notice to watch out for ASD cases that may be linked to vaccines. The British Parliament was already dealing with a national MMR crisis after pulling Pluserix off the market in Great Britain for causing outbreaks of viral meningitis. In other words, Wakefield and his coauthors did not create a link between MMR vaccine and autism. They only reported what others, including the IOM, were saying about a possible link between MMR vaccine and autism. In response to public concerns, the CDC has since funded studies on the issue, in which researchers concluded that MMR vaccines are not linked to autism.

The plan carried out by Parliament member Evan Harris, who escorted Brian Deer over to *The Lancet* and accused Dr. Wakefield and his coauthors of scientific fraud, worked. They succeeded in getting the scientific community and the world media to blame Wakefield for a global public health disaster created by government officials and the GMC's lawyers in league with the vaccine industry in the United States, Canada, Great Britain, and elsewhere throughout the world. It worked because everybody had their hands in the pie: top government officials, leading universities funded by the vaccine industry, and even prestigious scientific journals on the payroll of Merck and GSK—all searching for a scapegoat. What happened was exactly what President Eisenhower saw coming, and warned America about when he left office. But, given enough time, history has a way of righting itself. The question is: Where will it go from here? Will the common people around the world take up pitchforks and demand that scientific integrity be restored? Or, will government, industry and academia keep marching lockstep toward successive global public health and environmental disasters until the system finally runs out of people to blame and the resources it requires to recover?

With regard to the current public health crisis over MMR vaccines, here's what concerns me. Most of the viruses in cell cultures used to make live vaccines are probably attenuated, and, therefore, are less capable of replicating and causing infections. The processes used to denature these viruses (make them attenuated), however, are not 100 percent effective. Also, MMR vaccines, like all other vaccines, are never 100 percent pure. They're contaminated with low levels of other strains of measles and rubella viruses that may or may not be attenuated. My concern is that antibodies produced by MMR vaccines may keep measles and rubella viruses that escape the attenuation processes from causing outbreaks, but not prevent sporadic cases that go unreported. There's nothing to prevent these sporadic cases of measles and rubella infections from generating sporadic cases of autism just like those reported in the scientific literature. Sporadic cases can also plant the seeds of future pandemics involving atypical strains of measles and rubella.

Moreover, vaccines may not prevent the contaminant strains of measles and rubella viruses from generating large numbers of sub-clinical infections, which go unreported. Just as the CDC advertises with seasonal flu vaccines, even when vaccines fail to prevent infections, they can make the symptoms milder. Sometimes, the symptoms are so mild that they are not recognized as cases of measles, rubella, or whatever. Autism may be the least among the problems created by vaccines that go awry. But, still, widespread sporadic cases of measles and rubella viruses could produce a significant number of autism cases in which their links to MMR vaccines go unnoticed. To address these possibilities, studies that can detect even sporadic cases of measles and rubella viruses, which may be occurring by the mechanisms described above, should be conducted. These studies should be conducted in a manner in which the public can have full confidence in the results. If nothing else, these studies could help put to rest the issue of whether MMR vaccines can cause autism and, if so, what the actual risks are.

THE GMC FORMULA

CONFLATE THE EVIDENCE, CONDEMN THE MESSENGER

An email I received in 2011 from Professor Ian Booth, the General Medical Council's (GMC) expert pediatric gastroenterologist, shows that the GMC's solicitors instructed him to compare the Royal Free Hospital's routine pathology reports with Table 1 of the 1998 *Lancet* article by Wakefield *et al.* The purpose was to accuse Andrew Wakefield, and potentially Amar Dhillon and Andrew Anthony, of research fraud.[1] Until I knew that, it made no sense why Booth ignored Dhillon and Anthony's grading sheets, which could help clear up the issue of whether Wakefield fabricated any of the *Lancet* children's diagnoses. Brian Deer and the *BMJ* ultimately carried out the deed assigned to Booth when they published a series of articles that began in 2010 with "Wakefield's 'Autistic Enterocolitis' Under the Microscope."[2]

Before I could understand what really happened in Wakefield's case, I had to comb through transcripts of the GMC's hearings that lasted from July 16, 2007, until April 14, 2010. I also had to locate some of the GMC's original evidence, which is still retained in Wakefield's personal files. Then I had to track down additional documents from one of Wakefield's attorneys in England. Finally, I had to get Booth to clear up perhaps the most important part of the whole story, which only he and the GMC's solicitors could resolve. Knowing who was behind the analysis of the *Lancet* children's medical records Booth used in his report could explain how Deer came to publish the same analysis as his own work four years later.

Booth's expert report, which was submitted to the GMC on November 8, 2006, further illuminates this mystery. On its cover, it displays the following acknowledgment: "Prepared on the Instructions of: Field Fisher Waterhouse, 35 Vine Street, London". According to a letter to Brian Deer from Matthew Lohn, a Partner at Field Fisher Waterhouse (FFW), Deer was an informant for FFW when it handled the GMC's case against Wakefield and his coauthors.[3] In his letter dated May 25, 2005, Lohn clarified Mr. Deer's role and explained that he was not a *complainant*. (Complainants would be subject to cross-examination in the GMC's hearings.) Lohn wrote:

Dear Brian ... I write further to your telephone conversation with Peter Swain last Thursday seeking clarification in relation to your role in the above General Medical Council ("GMC") proceedings. ... As stated in Peter Swain's letter to you dated 16 December 2004, your role in this matter is that of 'informant' rather than 'complainant'. ... We are grateful for information supplied by you and your assistance to date.

It's easy to understand how Booth's comparison of Table 1 of the *Lancet* article with the children's medical records could have ended up in Deer's hands to publish as part of his own investigation in the *BMJ*. It's also clear why solicitors working for FFW and the GMC never introduced it as evidence in the GMC's proceedings. As pointed out before, it makes no sense to compare the *Lancet* children's routine pathology reports with the diagnoses in Table 1, instead of using Dhillon's blinded analyses recorded in his grading sheets—which is what Table 1 was based upon. Defense attorneys, I'm sure, would have taken Booth's analysis apart. It made far more sense to let Deer publish Booth's analysis as his own investigative report after the GMC's proceedings concluded.

The advantages were obvious. In Deer's hands, Booth's analysis of patient records would look like it was just part of an independent investigation by a news reporter. Brian Deer, its new "author"—who FFW made sure was an *informant*, not a *complainant*—wouldn't be subject to cross-examination at the GMC's hearings. Lancet editor Richard Horton, who was a witness for the GMC, could easily take care of retracting the Lancet article based on the GMC's findings regarding ethics violations. Then Brian Deer and the *BMJ* editors could take it from there. With the GMC proceedings behind them, a public lynching outside the courtroom over "new" allegations of research fraud would finally put the nail in Wakefield's coffin. As the BMJ editors put it: "The *Lancet* paper has of course been retracted, but for far narrower misconduct than is now apparent ... leaving the door open for those who want to continue to believe that the science, flawed though it always was, still stands. We hope that declaring the paper a fraud will close that door for good."[4]

Dr. Godlee and her coeditors at the BMJ concluded:[5]

... it has taken the diligent scepticism of one man, standing outside medicine and science, to show that the paper was in fact an elaborate fraud. Building on [Deer's initial] findings, the GMC launched its own proceedings that focused on whether the research was ethical. But while the disciplinary panel was examining the children's medical records in public, Deer compared them with what was published in the Lancet. His focus was now on whether the research was true.

Presenting Brian Deer as a brilliant investigative reporter who discovered an elaborate fraud by comparing the children's medical records with the *Lancet* article made a great story. But it wasn't true. Professor Ian Booth, who was also working on the case against Wakefield and his coauthors for Field Fisher Waterhouse, was the real man behind the curtain. He was the first person, but apparently not the last, that the GMC's lawyers instructed to compare the children's medical records with the *Lancet* article to create evidence Wakefield had faked the diagnosis of colitis.

What follows explains how it was all an elaborate illusion created by lawyers, scientists and editors with vested interests in defending the government's vaccination programs. For the *BMJ*, I believe protecting the MMR vaccine for its sponsors, Merck and GSK, was a driving force.[6] As best as I can tell, Fiona Godlee never disclosed this conflict of interest before publishing the grading sheets and other evidence I submitted. Earlier that year, she admitted before Parliament: "Even on the peer-reviewed side of things, it has been said that the journals are the marketing arm of the pharmaceutical industry. That is not untrue".[7] For Deer, his work with the GMC's lawyers was richly rewarded with two prestigious British Press Awards for prompting GMC hearings and the downfall of Andrew Wakefield.[8]

As it turns out, what appeared to be hopelessly complex actually conforms to a very simple formula, which the GMC's solicitors used to prosecute Wakefield and his coauthors. It is an extraordinarily effective weapon for suppressing research, and one that will undoubtedly be widely used by government, industry, and academic institutions in the future. It has two simple components: conflate the evidence and condemn the messengers. Below, I have outlined how this was used against Wakefield and his coauthors (Part A), and then me (Part B).

Part A: Defaming Wakefield and His Coauthors

— STEP ONE —

To condemn authors of the 1998 *Lancet* study for allegedly causing measles-related deaths with an "MMR scare," Deer and the *BMJ*:

- **Conflated** Wakefield's 1995 *Lancet* study that concluded that measles vaccines cause inflammatory bowel disease (IBD) with his 1998 *Lancet* study that concluded that IBD causes autism spectrum disorders (ASD), then claimed that the 1998 study concluded that MMR vaccines cause ASD.

  — Accusers worked for the *BMJ*, which is sponsored by manufacturers of MMR vaccines, giving the appearance that they were protecting their sponsors' commercial products. None of the accusers disclosed these conflicts of interest in their articles until I began submitting commentaries to leading science and medical journals with figures and references disclosing them.[9]

— STEP TWO —

In order to condemn Wakefield, Walker-Smith, and Murch for ethical misconduct, Deer and the *BMJ*:

- **Conflated** the research component of the 1998 *Lancet* study with a different study of twenty-five children with ASD and IBD, then claimed that the research was not covered under ethics approvals as the authors claimed in their article.

  — The research component (EPC 162-95) was approved by the Ethics Committee on Sept. 5, 1995, but the study of twenty-five children with ASD and IBD (EPC 172-96) wasn't approved until January 7, 1997, after collecting some biopsies.

— *STEP THREE* —

In order to condemn Wakefield for research misconduct, Deer and the *BMJ*:

- **Conflated** the Royal Free Hospital's routine pathology reports and other scientifically unreliable medical records to claim that Wakefield exaggerated the grading sheets.
  — To compare biopsy results with a table of diagnoses in the *Lancet* article, the accusers combined all of the scientifically unreliable medical records while completely ignoring the GMC's copies of the actual grading sheets Wakefield used to create the table, which came from an expert histopathologist using a standardized protocol in a blinded study with control samples.

Part B: Defaming Me

In order to claim that I was fired by EPA over research fraud, Deer and the *BMJ*:

- **Conflated** EPA documents related to a settlement agreement over my 1996 *Nature* commentary critical of EPA policies with an industry white paper falsely accusing me and my UGA coauthors of research fraud, while never mentioning that the allegations were withdrawn after EPA determined that they were false.

The reason that the conflate-and-condemn approach is so effective is twofold. First, it's extremely unlikely that any scientists will search out the original documents behind allegations of research misconduct and read them themselves. Second, unless someone has some specialized knowledge in the area, they won't understand the documents anyway.

Take Wakefield, for example. Thanks to Brian Deer and the *BMJ*, scientific and medical journals are filled with allegations of research fraud against him, except, in Wakefield's case, they're not called allegations. They're stated as facts and filled in anger, disgust, hatred, and dismay that he's still free to walk about on the face of the earth. Yet no academic institution, governmental body, or court of law has ever tried him, much less convicted him, of research fraud.

To summarize the allegations, Wakefield is accused of fabricating the diagnoses of colitis in eleven children with autism in order to create an "MMR scare" so that parents would stop using Merck and GSK's MMR vaccines. That, allegedly, is so that he could sell his own measles vaccine. Can you imagine what would happen if anyone in the world actually did such a thing, and ended up costing pharmaceutical companies billions of dollars in profits and killing thousands of children? Have you ever wondered why Merck and GSK haven't sued Andrew Wakefield, or why he has never been prosecuted for any of these heinous crimes against humanity?

It's odd, to say the least, that Brian Deer and the *BMJ* claim that they never saw Dhillon's and Anthony's grading sheets until I submitted them for publication. These are the documents they allege that Wakefield intentionally misinterpreted in order to fake the children's diagnoses of colitis. So we are to believe that Deer, an award-wining investigative reporter, never actually obtained any of the documents he says Wakefield used to create the "fake link" that created the global "MMR scare"? This is doubtful since Deer obtained copies of Wakefield's documents related to the case.[10]

Walker-Smith Exonerated

In March 2012, Justice John Mitting of the High Court of England dismissed all of the General Medical Counsel's (GMC) charges against Professor John Walker-Smith, senior coauthor of Andrew Wakefield's 1998 *Lancet* study.[11] And the GMC's counsel acknowledged that there was "serious weakness" in the panel's reasoning.[12]

Mitting criticized the GMC panel on numerous counts. He concluded, for example, that the panel made "fundamental errors,"[13] distorted evidence,[14] and based its findings on an inadequate analysis of the facts.[15] The panel provided inadequate and superficial reasoning and explanation for its conclusions,[16] inappropriately rejected evidence, relied upon "flawed" and "wrong" reasoning, and "numerous and significant inadequacies" in its conclusions, particularly in its findings in the individual cases of the *Lancet* children.[17]

In 2010, editors retracted the *Lancet* study based upon the GMC's determination that the patients were not consecutively referred and the study did not have ethical approval. The court, however, ruled that the *Lancet* children were, in fact, consecutively referred. It also found that medical procedures used in the study, including lumbar punctures and endoscopic biopsies, were clinically indicated and, therefore, did not require approval from an ethics committee. The High Court dismissed both charges upon which *Lancet* editors retracted the study; however, the journal has not reinstated the paper.

Although Justice Mitting corrected many of the GMC's errors, his approach to analyzing clinical studies was seriously flawed. He assessed, individually, whether each scientist's primary goal was to benefit the patients in the study versus patients in general. If the scientist's primary aim was to benefit the patients in the *Lancet* study, Mitting concluded that he did not need an ethics approval for his work on the project. Using this approach, Mitting concluded that Walker-Smith's primary goal was to diagnose and treat the *Lancet* patients, while Wakefield's purpose was "undoubtedly research."[18] Ethical approvals, however, are granted to *projects*, not *individuals*. They aren't like driver's licenses, in other words.

Ethics Statement Criticized

Mitting concluded that the following statement in the *Lancet* article was "untrue and should not have been included in the paper":[19]

LANCET *ARTICLE (1998)*

Ethical approval and consent Investigations were approved by the Ethical Practices Committee of the Royal Free Hospital NHS Trust, and parents gave informed consent.

The statement above has two parts: (1) the Ethical Practices Committee (EPC) approved the *Lancet* study, and (2) parents gave informed consent. According to a Freedom of Information (FOI) response from the London Strategic Health Authority of the National Health Service (NHS), Deer obtained copies of the ethics committee approvals covering the research component of the *Lancet* study in 2004.[20] These approvals are not mentioned in transcripts of the GMC hearings, and Dr. Wakefield confirmed that he was not aware that these records existed until *after* the hearings concluded. It appears, therefore, that Deer never disclosed these approvals to the GMC.

Absent these approvals, the appearance was created that Walker-Smith collected research biopsies from seven of the *Lancet* children without any ethics approvals. In other words, it made it appear that Wakefield and his coauthors were dishonest when they stated in the *Lancet* article, "Investigations were approved by the Ethical Practices Committee of the Royal Free Hospital NHS Trust, and parents gave informed consent."

Walker-Smith's attorney introduced three documents that clearly demonstrated that the research component to the *Lancet* study had been pre-approved by the Ethics Committee; the GMC simply disregarded them without cause. Judge Mitting stated:

> *The panel has heard that ethical approval had been sought and granted for other trials and it has been specifically suggested that Project 172-96 was never undertaken and that in fact, the Lancet twelve children's investigations were clinically indicated and the research parts of those clinically justified investigations were covered by Project 162-95 [the general permission given to Professor Walker-Smith in September 1995]. In the light of all the available evidence the panel rejected this proposition.*

To this, Justice Mitting stated:

> *Its conclusion that Professor Walker-Smith was guilty of serious professional misconduct in relation to the* Lancet *children was in part founded upon its conclusion that the investigations into them were carried out pursuant to Project 172-96. The only explanation given for that conclusion is that it was reached "in the light of all the available evidence.*
>
> *On any view, that was an inadequate explanation of the finding. As it may also have been reached upon the basis of two fundamental errors—that Professor Walker-Smith's intention was irrelevant and that it was not necessary to determine whether he had lied to the Ethics Committee—it is a determination which cannot stand unless it is justified by the detailed findings made in relation to the eleven relevant* Lancet *children.*

The records concerning EPC 162-95, to which Justice Mitting refers, were not nearly as comprehensive as the approvals Deer had obtained from

NHS. While Wakefield struggled to put all the facts together at the beginning of the hearings,[21] Deer apparently sat on most of the evidence Wakefield needed.

In the end, the GMC sanctioned Dr. Wakefield and Professor Walker-Smith for not complying with the "conditions for approval and the inclusion criteria" for Project 172-96 (i.e., investigating children before the approval date and failing to keep copies of the signed parental consent forms with EPC approvals for Project 172-96).[22] Dr. Wakefield and his senior coauthors, however, steadfastly maintained that EPC 172-96 had nothing to do with the *Lancet* study.[23]

Informed Consent

Regarding the issue of informed consent, the Court made no further comment. The record is clear, however, that Walker-Smith's group transmitted a report to the ethics committee titled "1999 Annual Report on Ethical Submissions 162-95 and 70-97."[24] It states, "Samples, with fully informed parental consent (using the consent forms as detailed in the submissions), were obtained from upper and lower endoscopies...."

I obtained the GMC's copies of the *Lancet* patients' signed consent forms from Wakefield's files. Patient 11, a US citizen, was not included. Based on other medical records obtained by the GMC, I listed the dates that colonoscopies were performed on each of *Lancet* children (see Table 1). All of the colonoscopies were performed after parental consent was granted on or before August 24, 1995.

"No Respectable Body"

Justice Mitting concluded, "There is now no respectable body of opinion which supports [Wakefield's] hypothesis, that MMR vaccine and autism/enterocolitis are causally linked."[25] It's difficult to believe that the High Court failed to understand that the absence of such a body of science may well be a direct, and intended, result of the GMC's actions against Wakefield and Walker-Smith. As mice vacate a field at the sight of an owl devouring even a single mouse, so it is with scientists. It's called the "ecology of fear."[26] If it were not for what was done to Wakefield and Walker-Smith, many outstanding scientists would likely be willing to conduct objective research

Table 1.   Lancet study signed consent forms[a]

Patient No.	Colonoscopy[b]	Reference[c]
01	07-21 to 07-26-96	Day 1, p. 10
02	09-01 to 09-09-96	Day 1, p. 08
03	09-08 to 09-13-96	Day 1, p. 11
04	09-29 to 10-04-96	Day 1, p. 12
05	12-01 to 12-06-96	Day 1, p. 17
06	~11-01-96[d]	Day 1, p. 14
07	01-26 to 02-??-97[e]	Day 1, p. 22
08	01-19 to 01-25-97	Day 1, p. 21
09	11-17 to 11-22-96	Day 1, p. 15
10	02-16 to 02-19-97	Day 1, p. 23
11	NA[f]	NA[f]
12	01-06 to 01-10-97	Day 1, p. 19

[a] Parental consents were obtained by August 24, 1995 (TA Reed & Co., Note 23)
[b] Royal Free Hospital admission to discharge dates (TA Reed & Co., Note 8)
[c] Transcript of GMC hearings (TA Reed & Co., Note 8)
[d] "On or about" November 01, 1996 (TA Reed & Co., Note 8)
[e] Discharge day in February 1997 was unrecorded (TA Reed & Co., Note 8)
[f] Patient 11, a US citizen, was not subject to the GMC investigations

on vaccine safety, and publish their results regardless of whether they may threaten government policies and industry practices. In response to letters to editor-in-chief Richard Horton calling for the reinstatement of the 1998 *Lancet* study by Wakefield et al., including from Dr. Wakefield, Dr. Horton stated, "We have no plans to change our decision about this paper."[27]

Over time, science has evolved into a sophisticated marketing tool for supporting government policies and industry practices. As mentioned earlier, even Fiona Godlee testified she would not disagree that peer-reviewed journals are "the marketing arm of the pharmaceutical industry."[28] Marketing has a lot more to do with hiding the truth than publishing, whether for promoting government policies or industry products. That's not to say that most science is untruthful. But, for the most part, the only areas of science that escape government manipulation are those that have little or no impact on government policies, guidelines, or regulations.

For now, the scientific community seems pleased with the outcome with regard to Andrew Wakefield, regardless of whether any or all of the allegations are true. But that's because there aren't many research scientists independently working on vaccine safety, and the fear of frightening millions of parents into not vaccinating their children far outweighs any concerns over comparatively small numbers of documented cases of adverse reactions to MMR vaccines. The same is true of research on biosolids. The number of researchers who are independently investigating adverse health effects in this could probably be counted on one hand, and the fear of water bills going up may keep most people from protesting over EPA and land grant universities covering up a few dead bodies here and there.

The problem is that, when government and industry are geared up to cover up, we have no way of knowing what the real toll is on public health or the environment. But one thing is safe to assume. Government agencies, big corporations, and the universities they fund will not confine their use of these tactics to attacking only scientists who represent small areas of interest. Nor will they hesitate to go after areas of science where they lack a broad consensus of support. Protecting government policies and industry practices knows no bounds. Whatever tools prove effective in reaching those goals will, sooner than later, be used with little if any restraint. It is something, as President Eisenhower warned, to be gravely regarded. Our silence now will eventually bring an unbearable price for all to pay.

GOVERNMENT DOMINATION OF SCIENCE

lthough corporate influence plays a major role in corrupting science, the federal government has a much larger impact because of the resources it controls and the power it wields. Imagine what big oil and pharmaceutical companies might do if they could levy taxes to build their infrastructure and imprison people who threaten their interests. Dwight D. Eisenhower, supreme commander of the Allied Forces in WWII and later president of the United States, could have warned future generations of many threats that loomed over the horizon as he left office. Yet, he regarded the industrial-military complex and the domination of science by the federal government to be two of the gravest threats the world faces.

Humanity depends on science overcoming the daunting technological challenges associated with billions of people living on a planet that couldn't have supported them for much of the past four and a half billion years of its existence. We are headed toward a day when the conditions that allowed human civilization to develop and flourish a mere twelve millennia ago will no longer exist. Diminishing natural resources, global warming, and the fouling of our water, air, and soil are accelerating its approach. Scientists, potentially, could find ways to get around at least some of these problems, or reverse their effects. But they are trapped in a system dominated by the federal government, which was created during WWII to support government and industry.

Much has been written about the corrupting effects that corporate influence has on science. But scientific journals at least require authors to disclose any commercial conflicts of interest. Most research on public health, the environment, and other important areas of science, however, is government-funded. I doubt that any editor has ever asked scientists funded by the federal government to explain how their research benefits the government agency that funded them. They know that scientists funded by corporations are strongly incentivized to support corporate interests, but seem to be completely unaware that these same forces operate within the government.

The Hatch Act and Anti-Lobbying Act prohibit federal employees from getting involved in partisan politics and lobbying Congress. And the Federal Grants and Cooperative Agreement Act (FGCA Act) prohibits federal agencies from using grants and cooperative agreements to directly benefit their programs. But, as we saw with EPA's National Biosolids Public Acceptance Campaign, that didn't stop the Office of Water from establishing a cooperative agreement with the Water Environment Federation in order to use congressional earmarks to support EPA's sewage sludge regulation and quash reports of adverse health effects.

A Few Bad Apples

At a conference on research ethics held at Harvard University in 2011, my presentation was titled "Institutional Research Misconduct."[1] It's a term I use to describe the manipulation of scientific research at institutional levels to support government policies and industry practices. Because the federal government is the largest institution involved in scientific research, it dominates the scientific endeavor as a whole.

The discussion that followed centered on whether government agencies, universities, and other institutions engaged in the manipulation of research to support government policies and industry practices involve just a few bad apples, or whether the institutions as a whole are corrupted. In the cases I've dealt with, the research misconduct is run by a few bad apples. It reminds me of Watergate. Institutional research misconduct often requires numerous employees to do their part, but only a few may know enough about the operation as a whole to realize that its mission is to support government policies and industry practices, destroy opponents by eliminating their funding sources, and cover up any illegal activity.

One example is the government's efforts to create a body of scientific research to promote the benefits and safety of biosolids. In this case, the unethical and illegal activities involve violations of environmental laws, including whistleblower provisions. The US Department of Labor, for example, ruled in multiple cases that EPA managers violated whistleblower provisions of multiple federal environmental laws by retaliating against me and other EPA scientists for publishing research results and other scholarly articles that raised public concerns about the safety of biosolids.[2]

By contrast, similar efforts by the *British Medical Journal* (*BMJ*) to defend its sponsors' vaccine products appear to involve most, if not all, of its managers. This includes its editors, its attorneys, and a reporter it used to publish allegations of ethics violations and research fraud. Here, the *BMJ*'s unethical behavior includes failing to disclose its financial conflicts of interest with manufacturers of the MMR vaccine when publishing editorials and articles by Brian Deer. In these writings, the editors and Deer accuse Dr. Wakefield of research fraud and ethics violations, including failing to disclose *his* alleged conflicts of interest with manufacturers of MMR vaccine. Moreover, falsely accusing someone of research misconduct is considered to be research misconduct itself in cases where the accusers know that the allegations are false.[3] Arguably, therefore, the *BMJ* and Brian Deer committed scientific fraud when they knowingly published false allegations of research fraud against me, and chose not to inform readers that a government body had found them to be false.[4]

With regard to biosolids, the same handful of scientists who created the program at USDA and EPA in the mid-1970s are still controlling it today, four decades later. Together, this small group has managed to redirect industrial pollution in the United States in the 1970s from water to land, and create a body of scientific literature to support the idea that the same pollutants that are harmful in air and water are beneficial when mixed with sewage sludge and applied to land. That's not to say that there isn't an army of industrial giants and powerful politicians working behind the scenes. But, still, internally within the federal government, all that was needed was a small, permanent contingency of scientists placed in the Office of Water (OW) and Office of Research & Development (ORD). One senior executive, Henry Longest, actively guarded the process for three decades. As a deputy assistant administrator for OW in 1978, Longest set the wheels in motion to create an industry-friendly sludge rule, which regulated no organic chemical pollutants and only a handful of heavy metals.[5] Once the final version of the rule was in place, he moved to ORD and managed the growing dissent among many of EPA's field scientists, who considered the rule as posing a serious threat to public health and the environment. During that period, Longest focused on shutting down my career in science.

When reporting in 2008 on issues involving my research, *Nature's* editors called EPA's biosolids program "an institutional failure spanning more than three decades—and presidential administrations of both parties."[6] To understand how a handful of federal employees can manipulate an area of science to support government policies, it may be informative to see how different parts of the US government failed to correct any of the abuses associated with EPA's biosolids program.

The White House

While Henry Longest was training EPA managers to shed their religious beliefs and sexual inhibitions toward coworkers, as discussed in chapter 6, the president of the United States awarded him the highest and second highest honors bestowed upon career senior executives.[7] Recipients are nominated by their agency heads and designated by the president for their integrity and commitment to excellence in public service, and receive the awards from the president in Rose Garden ceremonies. In 2013, President Obama halted the program, which doles out cash awards totaling 20 to 35 percent of the recipients' annual salaries, because of budgetary constraints.[8]

In 1996, I asked Jerry Melillo, associate director of the environment for the White House Office of Science & Technology Policy under President Clinton, if he ever spoke with EPA administrator Carol Browner about dealing with what was happening to ORD under Henry Longest. He said he went "around and around" with her over that, and was "overruled at the highest levels."[9]

Under President George W. Bush, a window of opportunity opened—momentarily. In 1999, when Bush was governor of Texas, he signed a bill providing incentives to lower pathogen levels in biosolids.[10] Local media at the time were covering widespread reports of adverse health effects from biosolids, which can contain a wide variety of pathogenic bacteria and viruses. John Howard, Bush's environmental policy director, met with me in Austin to discuss my research at UGA.

Later, when Bush was elected president, Mr. Howard called to tell me that the Bush administration would make solving the biosolids problem a high priority. In 2000, I met with Howard in the Eisenhower Executive Office Building, where presidents maintain an executive office and many of the White House staff work. As we talked, I handed him a short list of EPA officials, including Henry Longest, who had mismanaged the sludge issue over the years. I suggested that he find them a new place to work where they could no longer control it. He placed the list on his secretary's desk and said it would be taken care of the next morning.

Several weeks later, Dr. Russo showed me what's called a "turkey list," which came down through EPA channels. It contained the names I gave to Howard, plus others who were being moved out of ORD. It was the only time I ever had any hope that the White House would take care of the problem; it didn't last long. By 2001, the biosolids industry had figured out that I must have a contact inside the Bush administration. James Slaughter, an attorney representing Synagro Technologies, Inc., asked me during a deposition to identify my contact.[11] I have no idea what Synagro did with the information. All I know is that Longest and the rest of his rafter stayed in place, and I never heard from John Howard again.

In May 2004, as House Majority Whip Tom Delay boarded a private jet in Augusta, Georgia, Congressman Charlie Norwood handed him a document that Norwood had asked me to prepare. It was a summary related to various communications that executives of Synagro and the Water Environment Federation (WEF) were having with EPA administrator Christie Whitman during the 9/11 crisis. Synagro published its white paper containing false allegations of research fraud against me just days after the attacks on New York and Washington, DC. At the time, EPA headquarters had put me on a short list of national experts to contact in case a bioterrorism attack was also in the works.

As part of its efforts to deal with the crisis, EPA headquarters solicited bioterrorism-related proposals from selected scientists. Dr. Russo informed EPA headquarters that she felt that a project I proposed concerning certain combinations of bioterrorism agents was unique and important to national security. Syrian scientists had also contacted me about some of my research, and Norwood had forwarded their communication to the Department of Homeland Security. Coincidentally, *US News & World Report* later reported that terrorists in Iraq were working on a similar approach to the one described in my research proposal.[12]

Longest, however, nixed my proposal and terminated my employment in 2003. In taking this action, he ignored federal rules governing my assignment to UGA. Under the Intergovernmental Personnel Act, I was required to continue working at EPA for an amount of time equivalent to my UGA appointment, which made my earliest eligible retirement date May 2007.

Norwood was concerned about the biosolids industry pressuring the EPA administrator to investigate me for research misconduct during the 9/11 crisis. He felt that the situation warranted President Bush's personal attention, and Delay was headed directly to meet with the president.[13] Nothing ever came of the meeting, and, unfortunately, Norwood underwent a lung transplant several months later, and remained in declining health until passing away in 2007.

In 2011, *Nature* reported that the White House was reviewing nineteen scientific integrity policies submitted by federal agencies.[14] *Nature* reporter Eugenie Reich summarized my experience at EPA, in which Longest cut off my research funding, then offered to let me transfer to UGA for up to four years.[15]

> *Following complaints from a biosolids company, an EPA official used agency letterhead to spread allegations of research misconduct against Lewis, a situation that Lewis says led to him being forced out in 2003. A judge in 2007 dismissed Lewis's claims that he was fired illegally, but the EPA cleared Lewis of misconduct and reprimanded the official. With his science career nevertheless over, Lewis now works [with] the National Whistleblowers Center in Washington DC. Lewis's fate is the kind that a scientific-integrity policy released by the EPA on 5 August seems intended to prevent.*

In summary, the White House, at least in my case, was fully informed of EPA's retaliations from top to bottom over the course of several presidential administrations. President Eisenhower's warnings about government domination of science may still echo in some dark corner of the White House. But occupants of the Oval Office since his departure appear to be incognizant of their importance or completely impotent when it comes to acting upon them.

Congressional Oversight

In the fall of 1995, Representative John Linder, a dentist before running for Congress, stopped by my home as he made his rounds in Oconee County, Georgia. He offered to introduce me to Republican leaders focusing on EPA science, and arranged for me to meet with Newt Gingrich in the Speaker's office in Washington, DC. I took him up on the offer. While traveling back and forth to Washington, I also met with Charles E. Cooke, who worked for Texas Representative Ralph Hall. Hall was a Democrat serving on the Science Committee at the time. After switching parties in 2004, he became the committee's ranking member, then became chairman when Republicans regained control of the House in 2010. Cooke indicated at the time that many Democrats agreed that EPA had some serious problems when it came to integrating science in its regulatory framework. But they preferred to stand back, he said, and just let Republicans keep reinforcing their "anti-environment" reputation by trying to reform EPA.

I also approached Minority Whip Anthony Weiner of New York, a member of the House Science Committee. He had requested my help with infection control issues with flexible endoscopes, which he was working on with HHS Secretary Donna Shalala. Weiner and his staff, however, weren't interested in getting involved with problems associated with EPA science. So, I decided to continue working with Linder, so long as we picked an EPA regulation where science argues for strengthening, not weakening, it. It was the only way we could have any credibility. We focused on EPA's sludge rule, and our arrangement worked well for more than a year.

Things suddenly fell apart in March 1999, however, when Representative Linder arranged a press conference for me in the Rayburn House Office Building. Speakers were to include Linder and Representatives Richard Pombo and David McIntosh. They were to be followed by me, several other EPA scientists, and a special guest, Joanne Marshall. Joanne and her husband, Tom, at the time were involved in a lawsuit against Synagro over the death of their seventeen-year-old son, Shayne Conner. As described in chapter 3, residents in their New Hampshire neighborhood developed difficulty breathing whenever biosolids were applied to a hayfield at the end of their street. Shayne, who often suffered from the effects, stopped breathing and died in his sleep before he could be resuscitated.

Several days before the conference, someone from Linder's office informed me that Linder had met with other Republican leaders and decided that our conference shouldn't mention biosolids, and that Joanne Marshall shouldn't be included as a speaker. I replied that I wasn't dropping Joanne Marshall, and I moved the meeting to the National Press Club. A delegation of UGA administrators seeking congressional support met with Linder in his office in Washington not long afterwards. A member of the delegation told me later that, as soon as they entered Linder's office, Linder stood up and handed them a piece of paper with my name written on it. He said that Linder told them to go back and tell UGA president Michael Adams to "personally see to it" that I had no further contact with his office.

So, after being "banished from Woolworths," I decided to drive down to Augusta and visit Representative Charlie Norwood.[16] Normally, members of Congress can't cross over district lines to work with individuals outside their area without permission. But because I had been officially banned from my own congressman's office, Charlie felt that our forefathers would have wanted certain allowances to be made in such cases, "under the table," of course. He spent some time discussing a letter I drafted for members of Congress to send to Carol Browner regarding EPA's attacks on scientists and activists who were raising concerns about biosolids. Unfortunately, he was tied up writing the Patient's Bill of Rights, and couldn't see how it would be possible to help me until that was done.

I'll never forget what he said as I headed out the door: "David, the next time you're in Washington, stop by my office and I'll take you over to the House floor. You wait at a table outside, and I'll start grabbing members by the arm and bringing them out one by one. You take five minutes to tell them what you just told me, and I'll sit there nodding my head up and down. Then we'll see what happens." Even without the outfit, he was the closest thing to Santa Claus I've ever seen.

A few weeks later, I headed to Washington, DC, just in time to get ahead of the summer recess. With me, I carried handwritten letters from Alan Rubin, a scientist in EPA's Office of Water, threatening private citizens for raising public health concerns over biosolids. I also had an affidavit from the CEO of a biosolids company stating that Rubin had offered him "the pig business in North Carolina," if he would withdraw a paper he submitted, which discussed problems with pathogens in biosolids. To these documents, I attached a draft of a letter from members of Congress to EPA administrator Carol Browner asking her to explain Rubin's behavior.

Charlie sat me down at a little table outside the House floor, and began dragging members out during a break between votes. Billy Tauzin came out first, then Christopher Cox, followed by Don Young and others. Tauzin, who was about to become chairman of the Commerce Committee, said, "I think this is a message we can all get behind." Each member asked me what he could do. Before the House was called back into session, seven key members agreed to sign the letter.

Aloysius Hogan, who worked for Representative Joe Knollenberg, spent several weeks working with each member to get a version of the letter they were happy with. Everyone signed a committee version except Mr. Cox, who spent more than an hour revising it to get it like he wanted. Aloysius sent both letters to Browner, and I provided copies to Arnold Mann, a reporter at *Time* magazine. *Time* broke the story, and then followed with a second article about a CDC official, Joe Cocalis, who came out in support.[17]

The CDC gave Cocalis the job of putting together the first federal guidelines for protecting workers from exposure to pathogens when working with biosolids.[18] James Sensenbrenner, who chaired the House Science Committee, started the wheels turning for what was to become two full committee hearings into EPA's retaliations against scientists and activists over widespread reports of adverse health effects from biosolids.[19] Then, based on the Science Committee hearings, Congress passed the No Fear Act of 2002 to better protect federal whistleblowers from retaliation.[20]

Representative Knollenberg, who served on the Appropriations Committee, approached me about working with my local congressman, John Linder, to fund my research at UGA. I explained that I was banished from both UGA and Linder's office, but he wanted to try anyway. To say the least, Linder wasn't happy to see me back in his office. He commented that my research was "controversial," and that I should just talk with UGA about including another proposal in the stack UGA would be sending him. Yeah, right.

While I was in Linder's office, Sensenbrenner called twice, wanting me in his office to talk about the upcoming biosolids hearings. Linder wouldn't let me take the calls, but told me he would give me a ride over to Sensenbrenner's office after we talked privately. When Knollenberg's staff member left, Linder explained that he was concerned about farmers not being able to afford commercial fertilizers. I told him that if they keep loading up their farms with industrial wastes, their soils will eventually be too toxic to grow crops. The only good part about our meeting was leaving it. Really, it was the most interesting part of my day. John opened what looked like a closet door next to his desk. I assumed he wanted his overcoat. But it turned out to be a secret passageway down to a small parking lot where his car was parked. He gave me a ride over to Sensenbrenner's office, and said goodbye. That was the last time I ever saw or heard from him.

Despite the heroic efforts of Charlie Norwood, two congressional hearings, and the No Fear Act, nothing changed at EPA. In fact, it's only gotten worse. EPA no longer regulates composted biosolids, or requires that it be tracked or identified as biosolids. And, before passing the Senate, the Senior Executives Association (SEA) lobbied the committee to insert a "poison pill" in the No Fear Act (H.R.169) to protect senior executives, e.g., Henry Longest. SEA reported, "After months of work with Senate and House Staff, and members of the Government Affairs Committee in the Senate, much of the bad language has been deleted or substantially altered, and specific language has been inserted stating that managers would not be adversely affected by the bill."[21] The No Fear Act, therefore, did little if anything to stop federal managers from retaliating against research scientists who publish results that don't support government policies and certain industry practices promoted by government agencies.

Academia

The Federal Grants and Cooperative Agreement (FGCA) Act of 1977 prohibits the use of federal grants and cooperative agreements to directly benefit the government. Violations are punishable by fines and imprisonment. If academic institutions actually complied with this law, the government would find it more difficult to manipulate scientific research by funding universities to support its policies. But then that would mean turning down huge amounts of federal funding.

Once again, EPA's use of grants and cooperative agreements to fund land grant universities to support biosolids is a perfect example. The cooperative agreement that EPA's Office of Wastewater Management established with the Water Environment Federation (WEF) funneled millions of dollars in EPA funding and congressional earmarks to the wastewater industry's biggest trade association to support a National Biosolids Public Acceptance Campaign.[22] Internal EPA paperwork submitted to EPA's Grants Office stated that the project's objective was to gain acceptance of the science and the substance of EPA's sludge rule and overcome misinformation spread by opponents.[23]

In addition to prohibiting grants from directly benefiting the government, the FGCA Act requires that the recipients allow historically black colleges and other minority institutions and businesses to compete for any work contracted or subcontracted under the grants. For research on biosolids, this is a particularly good idea. Because minority communities are disproportionately impacted by land application of biosolids, historically black colleges should be less inclined to publish data that are fudged to cover up adverse health effects. But universities that produce "biosolids science" to support EPA's sludge rule are never historically black.

Eliot Epstein, for example, an adjunct professor at Boston University, who was funded by EPA's Office of Water, prolifically published articles dismissing health and environmental concerns attributed to biosolids. In 2001, he wrote to the head of BU's Department of Environmental Health to protest my being invited to speak. The letter, which the department head refused to act upon, stated:[24]

> *I feel that the way the conference is arranged will diminish Boston University School of Public Health's credibility. The selection of speakers is extremely poor.... Dr. Lewis has absolutely no standing in the scientific community in this area.... Dr. John Walker or Bob Bastian of EPA would have been much more credible than David Lewis....*
>
> *In the future, if the Boston University School of Public Health is interested in seeking funds for research on health aspects on this subject, it must have the reputation of a credible scientific institution. There are numerous sources of funds for research available. The University of New Hampshire, Pittsburg University, Tulane University, University of Arizona, Johns Hopkins University, and numerous others have received considerable grants on this subject.*

After the dairy farmers and I filed a qui tam lawsuit over EPA and UGA employees using a federal grant to publish fabricated data, our attorneys deposed UGA administrators and faculty members involved in the study. The former vice president for research, who certified that UGA would comply with the FGCA Act, downplayed his authorization. FGCA Act violations involving EPA contracts and cooperative agreements approved by this same UGA official were the subject of a high-profile EPA Office of Inspector General investigation and a hearing before Congress in 1993.[25] He said someone else signed for him, and that no one in his office was expected to actually read the grant applications.[26]

Judge Clay Land of the Federal District Court in Athens, Georgia, eventually dismissed our qui tam complaint on the grounds that my attorneys obtained a copy of the grant application under the Georgia Open Records Act. In other words, the court did not consider the dairy farmers and me to be the original source of any information showing that EPA and UGA employees intentionally used an EPA grant to publish fabricated data. Never mind that I could have obtained the grant application myself as part of my official EPA duties when I was assigned to UGA And never mind that the dairy farmers were the first to discover that the data were fabricated, and that I had discovered the purpose of the grant from my independent investigations at UGA. Altogether, UGA spent over $250,000 paying private law firms to keep from having to withdraw the fabricated data.[27]

As of 1997, about one-third of all research expenditures at land grant universities came from federal grants.[28] In 2003, the provost and other UGA administrators advised the chairman of the Department of Marine Sciences not to hire me because EPA and the wastewater industry were out to get me. One UGA administrator explained:[29]

> *We're dependent on this money...grant and contract money... money either from possible future EPA grants or* [from] *connections there might be between the waste-disposal community* [and] *members of faculty at the university.*

If UGA is any indication, a number of universities have no intention of complying with the FGCA Act, and will go to great lengths to protect any data they publish to support government policies.

10

INSTITUTIONAL RESEARCH MISCONDUCT

When I was an undergraduate student, I worked part-time at the EPA lab in Athens, Georgia, to help cover my college expenses. Pat Kerr, a biologist in my branch, was locked in a dispute with EPA headquarters over the issue of whether phosphates in laundry detergents caused algae to proliferate in lakes and rivers. Algal growth is limited by carbon, not phosphorus, she argued. Her opinion took center stage late one night when Johnny Carson had Eddie Albert, the star of the hit TV series *Green Acres*, as a guest. While discussing his interest in environmental issues, Albert commented about phosphates and algal blooms, as they're called, then added, "But Dr. Kerr at EPA thinks that phosphorus doesn't stimulate algae." Pat's funding in Athens never suffered and, in the end, she turned out to be right. Phosphorus stimulates bacteria to produce more CO_2, which, in turn, causes algae to proliferate.

Tolerating dissent began to die out in the mid-1990s as science at EPA and other federal agencies became more politicized. I remember joking about it with CDC scientists when I was working on infection control in dentistry. To them, it was no laughing matter. They were packing up and leaving.[1]

Although it didn't make headlines, a similar exodus occurred at EPA during the mid-1990s. One leading EPA scientist remarked to me that he was so embarrassed about working at EPA that he just referred to his adjunct professorship when other scientists asked where he worked. It was all about political control from the top down.[2] At the same time, a survey of 1,600 scientists in EPA's Office of Research & Development (ORD) by the Union of Concerned Scientists found that over half of them had experienced political interference within the previous five years.[3] Twenty-two percent witnessed "selective or incomplete use of data to justify a specific regulatory outcome," and 17 percent were "directed to inappropriately exclude or alter technical information."[4] Many admitted that they feared retaliation if they failed to comply.

Since 1993, the manipulation of research at the federal level to support government policies and industry practices has become one of the greatest threats to public health and the environment. If scientists inside and outside of the federal government don't push back, the future of science in any area in which the government has a stake is very bleak. But overcoming a large body of scientific literature that has been manipulated by government and industry is no small challenge. Doing it without losing your job, or landing in prison, is even more challenging.

Role of Activists

In July 2013, the National Institute of Environmental Health Sciences held a conference on environmental health disparities and environmental justice.[5] Cosponsored by EPA, the CDC, the National Institute on Minority Health and Health Disparities, the Office of Minority Health, and the Indian Health Service, the organizers sought to prioritize research and identify new environmental exposures that could lead to health disparities. Steven Wing of the University of North Carolina at Chapel Hill invited me to speak on efforts by polluters to silence research scientists and local activists. Activists attending the session spoke of being shot at, stopped along the highways, and losing their jobs. It is a part of the Civil Rights struggle in the United States that is still rampant and stuck in the 1960s.

As was evident in the Science Committee hearings in 2000, federal agencies target activists as well as scientists who attempt to document problems associated with government programs.[6] In every area where I have addressed government policies, concerned citizens played an important role in the research that my coauthors and I published. Our investigations into adverse health effects associated with biosolids, for example, could never have succeeded without information provided by activists impacted by land-application practices. Although our peer-reviewed research was airbrushed from the 2002 National Academy of Sciences report in order to conclude that no cases of adverse effects were documented in the scientific literature, it still cited hundreds of anecdotal cases compiled by Helane Shields.[7] Those cases provided the information we needed to design our studies.

Abby Rockefeller, who pioneered efforts to deal with EPA's policies on sewage sludge, had a huge impact on my research.[8] In 2001, she organized a conference on the subject at Boston University and invited me as one of the speakers. Caroline Snyder, another pioneer in this area, has doggedly stayed on top of misinformation published by government, industry, and the scientists they fund to support EPA's sludge rule.[9] She had an enormous impact on our research. Similarly, Nancy Holt's efforts to correct the environmental injustices associated with land application of treated sewage sludge in North Carolina was pivotal in Steve Wing's research in this area. Community activists, therefore, play a vital role in detecting and combating institutional research misconduct.

Curing Institutional Research Misconduct

After finishing law school this year, our youngest son, Jedd, plans to begin his Ph.D. program in biology at Stanford University. To prepare, he spent a couple of extra years taking science courses and working in a cancer research lab at Harvard University. When I asked him to share his thoughts on institutional research misconduct, he likened institutions to the human body, and suggested that it be treated as an "organizational disease."[10]

> *Methods used to fight diseases of the body include prevention, detection, and treatment. The same thinking can be applied to combating organizational corruption. It could be **prevented** by minimizing incentives to commit fraud, and the ability to collude, e.g., by constructing an organizational firewall between EPA's ORD labs and program offices to keep policymakers from dictating which research is done, or more importantly, not done. It could be **detected** by mandating government transparency, and **treated** with laws that require investigating suspected or alleged institutional-level research misconduct, and punishing wrongdoers.*

He cautions, however, that any laws created to combat institutional research misconduct can and will be used by the institutions to discredit those who uncover the fraud. Moreover, the deck will always be stacked in favor of fraudsters supported by credible organizations, which have far greater resources, and usually more political savvy, than individual scientists. They could easily win any battle over credibility hands down. In my own case, I concentrated on government policies that blatantly defy common sense. Hopefully, anyone with even a little common sense would believe me over government officials and their government-funded research. People understand that the scientific literature is being constantly revised. And, so that anyone could understand the issues, I used simple visual demonstrations.

With dental infection control, for example, I had producers at ABC's *Primetime Live* film my demonstration of visible amounts of bright red blood leaking back out of dental drills and prophy angles after a dentist prepared them for the next patient according to CDC guidelines. With flexible endoscopes, I used photographs showing how peracetic acid, which is used in about 20 percent of the hospitals in the United States, can restore internal channels coated with feces and other patient materials back to their original white color.[11] With biosolids, I demonstrated how hundreds of thousands of pathogens on Petri plates can be seen growing from biosolids treated according to EPA standards.[12] Government agencies will always win when the outcome rests upon which side has the most scientific articles to support its arguments. Whenever honest scientists try to overcome a large body of dubious scientific literature, they need something more compelling than a few scientific articles to win the battle.

Another thing I had going for me was the contacts I had with reporters at major news media outlets in the United States, and with key members of Congress. When EPA headquarters sent out emails accusing me of violating government ethics rules and the Hatch Act with my *Nature* commentary, leaders in the House and Senate sent letters to EPA administrators.[13] At the same time, *Time* magazine, *USA Today*, the *Washington Post*, and other major news media outlets relentlessly hammered EPA over its sludge rule and retaliations against me.[14] National media coverage of other research my coauthors and I published was also vital to influencing change within government and industry.[15]

EPA settled all but one of my whistleblower complaints before hearings could even be scheduled. And, if they had settled my last Labor Department case and qui tam lawsuit the same way, I would have never had enough discovery documents to write this book. By contrast, I know from working with the National Whistleblowers Center that scientists who challenge government policies are almost always caught by surprise when retaliations come. They enter the battle like foot-soldiers going off to war with no training, no advance preparation, no weapons, and no idea of what to expect or how to deal with it. They don't stand a chance.

Government Conflicts of Interest

Federal grants and cooperative agreements are the primary instruments that the government uses to build a body of scientific literature to support its policies. Unfortunately, all three branches of government have failed to enforce the Federal Grants and Cooperative Agreement Act, which expressly prohibits federal agencies from using these instruments to directly support the government. And it's probably asking too much to expect academic institutions to comply with a law that the Justice Department has shown no interest in enforcing, especially when a large portion of their research funding comes from the government.

At a minimum, however, scientific journals should start requiring authors to declare any financial conflicts of interests associated with government funding. For example, authors submitting an EPA-funded study on biosolids concluding that biosolids are safe and environmentally beneficial should include a disclosure stating, "This research was funded by EPA, which has a policy that biosolids are safe and environmentally beneficial." Likewise, authors of a CDC-funded study declaring that the HPV vaccine is safe should include a disclaimer stating that the CDC recommends vaccination against HPV.

Corporate Funding

In 1993, Steris Corporation in Mentor, Ohio, invited me to meet with corporate executives and its scientists and engineers who developed an automated process for using buffered peracetic acid to sterilize flexible endoscopes. The devices aren't manufactured to withstand sterilization. It was about a $1-billion business and rapidly growing. But the CDC was pressuring the FDA to withdraw the company's sterilization claims because it's not possible to sterilize a medical device using liquid chemical germicides, if the device is grossly contaminated with patient materials and impossible to clean. In one of the studies I published, I proved this to be the case based on the diffusion rates of germicides under the conditions they are used to clean and disinfect endoscopes.[16]

During the meeting, I tried to explain why the data Steris Corporation submitted to the FDA were insufficient. The problem was that their process sterilized only the outside of tiny bits of tissue and layers of biofilms trapped inside endoscopes. Once these particles dislodged and entered other patients, they could break down and release infectious viruses and bacteria. The methods Steris used to test for sterility were not designed to detect this problem. As time dragged on, our arguments grew more heated. At one point, the president of the company, Bill Sanford, was literally shouting at me at the top of his voice. But then, as the scientist who developed the process stared at a sketch I made on a chalkboard, a light turned on. He looked at me and said, "I see what you are saying." For several minutes that followed, you could have heard a pin drop.

I looked at Bill, and said there was one experiment he could do that may support the company's sterilization claims. Most endoscopes are disinfected with glutaraldehyde. Like formaldehyde, which is used for embalming, glutaraldehyde causes proteins to bond together. That makes them much harder to remove from inside tiny channels and other internal areas of flexible endoscopes. I pointed out that peracetic acid is used to clean cross-linked proteins and other hardened organic matter out of narrow channels in laboratory glassware. It may work the same way in endoscopes, and, if so, it could actually sterilize the surfaces in the process.

I explained to Bill that if this study proves that the Steris process *removes* layers of patient materials hardened with glutaraldehyde from the internal surfaces of endoscope channels, the company's sterilization claims were valid so far as I was concerned. But, if the process failed, the FDA would likely use the results to withdraw the company's sterilization claims. Bill thought about it for about ten to fifteen seconds, then stood up and said, "I don't care how it turns out; just do it." Then he walked out of the room without saying another word.

Corporate funding doesn't necessarily rob research of its credibility. In this case, the company's president decided that the company would stand or fall based on the outcome of the research, whichever way it turned out. The company's senior vice president later asked if I would do the research. When I declined, he found a group at Case Western Reserve University's Department of Biomedical Engineering to conduct the study. Fortunately for Steris Corporation, the results supported its sterilization claims. The CDC still uses the study to support its current guidelines for disinfection and sterilization in health-care facilities, which includes sterilization of flexible endoscopes with peracetic acid.[17]

When I was assigned to UGA to await retirement, Henry Longest required that I obtain all of my research funding from sources other than EPA. I turned down offers from Synagro Technologies, Inc., to fund my biosolids research, and funded that part of my research out of my personal savings. My assignment agreement, however, required that I also conduct research on medical and dental infection control. I wasn't about to take a chance on EPA having the Justice Department prosecute me for violating the terms of my settlement agreement. So I talked with some of the corporate executives I knew and trusted in the infection control business.

Vision Sciences, Inc., which manufactures sheaths for flexible endoscopes, funded my prospective epidemiological study of endoscope-related infections in Egypt. And the National Association of Dental Laboratories (NADL) funded a study I designed on risks of HIV associated with handling dental molds used to make crowns and bridges. To carry out these projects, I worked with the University of Maryland's Department of Epidemiology and Washington University's Retrovirus Clinic.[18]

Vision Sciences stood to profit from my study in Egypt if it demonstrated that current guidelines for disinfecting flexible endoscopes failed to protect patients from cross-infection with hepatitis C. The NADL was pushing for government-mandated infection control standards for dental laboratories, where workers have to use sharp instruments to cut through dental molds containing pockets of blood. Our results were interesting and unexpected. In Egypt, only two out of 149 at-risk patients appeared to contract HCV (sero-converted), but follow-up tests were unable to detect the virus in blood samples. Both patients cleared the virus, or the initial test results were false positives.[19]

The problem with infection control in endoscopy is not so much the guidelines as it is compliance. One national survey found that two out of three physicians have their flexible endoscopes cleaned for the next patient in considerably less time than is required to perform all of the steps recommended in infection control guidelines. That's one reason I recommend having your colonoscopies done at a clinic using automatic washer-disinfectors or, preferably, automated sterilization using buffered peracetic acid.

In our dental study, we found that all of the polymers used to make dental molds contain one or more substances that instantaneously render HIV noninfectious. In the end, neither Vision Sciences nor the NADL benefited financially from our results. Thus, these three cases, involving Steris Corporation, Vision Sciences, and the NADL, demonstrate two important lessons. Corporations aren't always looking to manipulate research, and it's the character of the researchers and quality of their research that determine whether their work is credible, not their source of funding.

11

THE NUREMBERG CODE

n 2001, the Maryland Court of Appeals likened an EPA-funded study, which Mark Farfel and others at the Johns Hopkins Kennedy Krieger Institute (KKI) performed on low-income families in Baltimore, to Nazi war crimes.[1] The Court cited the Nuremberg Code (see Appendix VII: The Nuremberg Code) and the Declaration of Helsinki crafted by the international medical profession after WWII, and concluded, "Serious questions arise in this case under either code."[2] In order to approve the study, Johns Hopkins University "encouraged the researchers to misrepresent the purpose of the research in order to bring the study under...a lower safety standard of regulation." Plaintiffs, including a small child, developed elevated blood-lead levels associated with lead poisoning.[3]

Several years later, EPA, USDA, and the Department of Housing and Urban Development (HUD) funded Farfel and his team to perform similar experiments on an African American community in Baltimore.[4] There, biosolids were contaminated with lead levels that posed an increased risk of lead poisoning once they were mixed with the neighborhood's lead-contaminated soil. The government also funded some of these same researchers to feed prisoners capsules containing biosolids and soil collected from a hazardous waste site. Based on these experiments, EPA and USDA claim that application of biosolids in inner-city communities protects children from lead poisoning.

The Maryland Court of Appeals elaborated:[5]

Why wasn't the Nuremberg Code immediately adopted by
United States courts as setting the minimum standard of care for

human experimentation? One reason, perhaps, is that there was little opportunity. As remains true today, almost no experiments resulted in lawsuits in the 1940s, 50s, and 60s. A second reason may be that the Nazi experiments were considered so extreme as to be seen as irrelevant to the United States. This may explain why our own use of prisoners, the institutionalized retarded, and the mentally ill to test malaria treatments during World War II was generally hailed as positive, making the war "everyone's war." Likewise, in the late 1940s and early 1950s, the testing of new polio vaccines on institutionalized mentally retarded children was considered appropriate. Utilitarianism was the ethic of the day. . . . Noting that the Code applied primarily to the type of outrageous nontherapeutic experiments conducted during the war, physician groups tended to find the Code too "legalistic" and irrelevant to their therapeutic experiments, and set about to develop an alternative code to guide medical researchers. The most successful and influential has been the World Medical Association's (WMA) Declaration of Helsinki . . .

The Declaration of Helsinki was crafted by the international medical profession, as preferable to the Nuremberg Code crafted by lawyers and judges and adopted right after the Second World War. The Declaration, or, for that matter, the Nuremberg Code, has never been formally adopted by the relevant governmental entities, although the Nuremberg Code was intended to apply universally. The medical profession, and its ancillary research organs, felt that the Nuremberg Code was too restrictive because of its origins from the Nazi horrors of that era.

Recollections of the Third Reich

Caroline Snyder, who wrote the afterword to this book, was among the millions of German citizens who escaped or were expelled from Germany toward the end of WWII. Her comments on the subject are included below.[6]

As a former East German who became a naturalized US citizen, I remember our midnight flight from the approaching Red

THE NUREMBERG CODE   171

Army in February 1945, shortly before WWII ended. My father, who was killed in 1942 on the Russian Front, had inherited and managed several large agricultural estates. We children learned early on about sustainable farming practices and the value of land. During the war there was little fuel to run farm machinery. Even children worked on the fields, pulling flax, collecting potato beetles in glass jars, and learning how to manage draft horses. Large farms in this region still grew a variety of food, feed, and fiber crops, integrated with livestock production. It was a system emphasizing stable yields, and protecting healthy soil for future generations in contrast to chemical-intensive CAFOs and monocropping, which are aimed at producing short-term high yields.

We fled westward a few days before the Red Army confiscated our land. Owners of large agricultural estates, who were perceived as capitalists accumulating wealth by exploiting workers, were a prime target of the conquering army. If we had not fled, we would have been deported, raped, or murdered as were many others. We were fortunate to stay with various relatives who provided temporary food and shelter until we obtained visas and passage to emigrate to the United States. We were a few of the fourteen million Germans who either fled or were expelled from their ancestral homeland in Pomerania, Silesia, and Prussia.

In America, government and corporate control of research and the resulting erosion of democratic principles have always caught my interest. I have often wondered about the German scientists employed by the Nazis. Did any of them ever speak out against Hitler's totalitarian government or join the resistance movement? Many religious leaders did. For example, members of the Confessing Church courageously challenged the establishment of a Nazified Protestant Church because it blatantly contradicted basic Christian principles. They also secretly saved hundreds of Jewish children from the gas chambers. Many who were active in the resistance movement were jailed, shot, hanged, or sent to concentration camps.[7]

So far, I have not yet found one German scientist who joined the resistance. Several of Wernher von Braun's group of rocket scientists, responsible for designing and producing V-2 rockets, held high SS positions. They were well-informed of the Nazi atrocities; some were actually in charge of the very agencies that directed medical experiments in the death camps. Yet, as best I can determine, none of them ever spoke out or resigned their position. They were aware that their weapons expertise was a valuable commodity, and used it to their advantage. Even before the fighting stopped, the US and Soviet Union began to recruit them for each country's space program. After weighing their options, the von Braun group decided to surrender to the US Armed Forces where they were welcomed with open arms.

While we had barely escaped and survived the onslaught of the conquering Russian army, the rest of the expellees were not so fortunate. Of the fourteen million who were forced to leave, two million perished. During the last months of the war, thousands froze to death or drowned when their rescue ships were sunk by Russian torpedoes. Many thousands died in the Dresden fire bombing. Those, unable to escape in time, were repeatedly strafed, raped, starved, or killed by Russian soldiers. Because of the agreements reached at the Potsdam and Yalta Conferences and the Morgenthau Plan, the free world stood by or even sanctioned these horrors. In the eyes of many, it was only right that the expellees, together with the rest of the post-war German and Dutch civilians, should be punished for the atrocities committed by the Nazis.[8]

Meanwhile, the rocket experts and other German scientists who actually took part in these atrocities were given special treatment and never held accountable. Since many of them were potential or actual war criminals, they had previously been identified as "a menace to the security of the Allied Forces." Truman insisted that any Nazi scientist actively involved in Hitler's military operations should not be hired for our space program. To get around this, the Joint Intelligence Objectives Agency established

Operation Paperclip.⁹ These scientists were given entirely new dossiers, completely erasing any information about their Nazi affiliations and activities. With all of the incriminating evidence officially expunged from the public record, they were able to obtain security clearances, become US government scientists, and, later, US citizens.

Other high-ranking German scientists, such as Kurt Blome, were also protected by Operation Paperclip. Blome specialized in chemical and biological warfare, conducted medical experiments on Dachau prisoners, including children. After he was acquitted at Nuremberg, and his past activities covered up, the US Army Chemical Corps hired him to work on the US chemical weapons program.⁹ Here, under Operation Paperclip, the US government enticed thousands of US soldiers to undergo potentially dangerous non-therapeutic medical experiments."¹⁰

A year after abandoning our agricultural estates in East Germany, we boarded the Gripsholm and arrived in New York Harbor with hundreds of other war refugees. As for so many immigrants before us, the sight of the Statue of Liberty evoked strong emotions among those on deck. Many wept openly. Some of us were being welcomed home. Others were getting their first glimpse of the free world. Perhaps those who have been deprived of basic human rights appreciate the true value of democracy the most. The Statue of Liberty, bathed in the light of the rising sun, was not an empty symbol. It represented hope and freedom where citizens could openly debate and question government policies without retaliation.

While teaching at the Rochester Institute of Technology, I learned firsthand what it meant to openly and honestly debate, write about, discuss, and question our government policies without retaliation. Knowledge thrives in an atmosphere of academic freedom, and the free interchange of ideas. Since then, honest research has become increasingly scarce as government agencies and big corporations levy their economic power to bring scientists in line with their own interests.

*It threatens democracy and human welfare when the very gov-
ernment agencies that are charged with protecting public health,
agriculture, and the environment work with the industries they
regulate in order to undermine that protection for their own
short-term gain. Government officials who actually lead the
assault on their own scientists who raise concerns have no place
in a democratic society. Their tactics are reminiscent of those used
by totalitarian regimes to silence dissidents.*

*The tactics used by those who oversee EPA's deeply flawed bio-
solids program are a prime example. They target economically
and educationally disadvantaged communities and minori-
ties for disposal of the nation's municipal and industrial wastes
in sewage sludges. They use children and adults living in these
communities for non-therapeutic medical experiments aimed at
determining the effects of hazardous wastes on human health.
They turn a deaf ear to pleas of help from those who suffer the
harmful effects of spreading chemical and biological wastes
around their homes, schools and churches. They use mass media
propaganda such as the National Biosolids Public Acceptance
Campaign to convince the population that toxic wastes applied
to land are completely safe and even beneficial. They stifle honest
public debate. These are all similar to tactics the Third Reich used
to control the German population and suppress the truth.*

Relevance of the Nuremberg Code

Caroline has often remarked to me that she sees parallels between what's
happening to dissenters in the United States, especially scientists, and what
her family witnessed in Germany as Hitler rose to power and his SS began
to terrorize the population. Before my father passed away in 1993, he told
me something I thought I would never hear from a Navy pilot who fought in
WWII: "Russia is becoming more like the United States," he said, "and the
United States is becoming more like Russia."

It's something I wonder about whenever I think back to a witch hunt
that EPA administrator Carol Browner and John Martin, the agency's
inspector general, subjected EPA's research scientists to in the early 1990s.[11]

Armed federal agents from Washington, DC, spread out across the country, knocking on the scientists' doors late at night, and prosecuting them with trumped-up charges.[12] It was a turning point for science at EPA. Before, scientists were free to publish results with little if any interference from EPA headquarters; afterward, that was no longer the case. Many of EPA's best scientists left. It will never be the same again. It was this nightmare that initially drove me to confront EPA headquarters, first with a *Nature* commentary—"EPA Science: Casualty of Election Politics"—and then a series of lawsuits that lasted 15 years. I had to find out for myself how this could happen in the United States, and where it was all headed. After reading my *Nature* commentary, Speaker Gingrich looked at me in his office overlooking the National Mall in 1993 and said, "You know you're going to be fired for this, don't you?" "I'm willing to be fired," I replied, "but I hope to stay out of prison." The Speaker laughed, and turned to a staff member and said, "I like this guy." But I wasn't joking. Only EPA scientists who experienced what we went through could understand how I was feeling. It leaves a lasting impression when armed federal agents from Washington, DC, start rounding up innocent scientists en masse and threatening them with prison terms.

Caroline talks about the role some of the German scientists who joined America's space program played in Nazi atrocities. Their treatment in the United States echoes the haunting questions that the Maryland Court of Appeals grappled with. Specifically, the Court questioned how we can differentiate the experiments EPA and other government agencies in the United States are performing on economically and educationally disadvantaged citizens from the experiments carried out in Nazi Germany.

One thing that weighed heavily in the Court's decision was the fact that Johns Hopkins University asked its researchers to mischaracterize the trials as beneficial to the test subjects so that they could intentionally expose them to harmful chemicals. So did the fact that the researchers chose not to inform the test subjects of the risks they were being subjected to, and even withheld the results of their experiments from the test subjects when their blood-lead levels began to rise. "The conflicts are inherent," the Court stated. "This would be especially so when science and private industry collaborate in search of material gains."[13]

The questions posed by the Maryland Court of Appeals go well beyond just the experiments EPA and Johns Hopkins conducted in Baltimore.

When the government employs such methods, which the Court likened to Nazi war crimes, to justify subjecting the general population to the same risks, the numbers of victims increase. Consider, for example, the lead abatement experiments EPA and Johns Hopkins University performed on an African American community in Baltimore several years later to justify spreading biosolids on lead-contaminated soils in inner-city neighborhoods. I've seen large numbers of children and adults covered with boils and breathing through respirators after being exposed to dusts from biosolids applied to land. Some have died when their conditions progressed to staphylococcal septicemia and/or staphylococcal pneumonia. While visiting land application sites where these problems are occurring, I've spent hours gasping for air, and have sat propped up on pillows all night as fluids solidified in my airways and obstructed my breathing. Knowing that government and industry have manipulated science, exposed mostly educationally and economically disadvantaged populations to the greatest risks, and covered up illnesses and deaths, how are the innocent victims of EPA's biosolids program any different from the families harmed by the experiments addressed by the Maryland Court of Appeals?

In my opinion, they can all be likened to Mengele's experiments for the same reasons that the Court drew that analogy. In the experimental trials, EPA and Johns Hopkins didn't inform the test subjects of the actual risks involved; then they failed to inform the test subjects of the results obtained from their blood samples, which indicated that the experiment was causing increased lead levels in their blood. When EPA and USDA use such experiments as this to justify subjecting the public to the very same risks, they repeat the same process by downplaying the risks and covering up any adverse health effects all over again. They even work with industry and academic institutions to silence scientists such as myself, who may document adverse health effects so that the public is made aware of the risks.

While working at EPA and collaborating with the CDC, I watched firsthand as both agencies politicized science and partnered with industry in ways that I, and none of my colleagues, had ever seen before. It began during the first Clinton administration and culminated during George W. Bush's second term with a mass exodus of the government's most experienced public health scientists. Five of the CDC's six former directors wrote a letter to then–CDC director Julie Gerberding, who now serves as president of Merck's vaccine unit.[14] They complained that "strategic shifts in the agency's focus are putting public health at risk." According to a spokesperson for Gerberding, she wanted to "focus more on leadership and less on operations."[15]

A similar exodus in Canada has been underway since 2006, as the Canadian government lays off scientists and expands its communication staff.[16] In 2013, more than four thousand federal scientists in forty federal departments and agencies dealing with public health and the environment completed a survey sponsored by the Professional Institute of the Public Service in Canada (PIPSC).[17] The overwhelming majority 90 percent responded that they are not allowed to speak freely to the media about their work; and almost 86 percent believe they would face censure or other retaliation if they spoke out about a departmental decision that could harm public health, safety, or the environment."

As mentioned earlier, EPA administrator Carol Browner and the Agency's inspector general literally terrorized scientists throughout EPA's dozen or so research laboratories across the country, causing a similar exodus of the agency's most accomplished scientists. Armed federal agents from Washington, DC, spread out across the country, knocking on the scientists' doors late at night, and shoving them into their chairs at work.[18] They threatened them with incarceration in federal prisons if they refused to wear hidden microphones to spy on lab directors and other senior government officials. None of our scientists cooperated, and a number of them were prosecuted.

Upon leaving EPA, Browner joined Madeleine Albright and other senior government officials to form the Albright Group, an international consulting firm. In 2008, president-elect Barack Obama tapped Browner as the White House coordinator of energy and climate policy. John Broder of the *New York Times* looked into the Albright Group at the time. The only clients the Group would disclose were Merck and Coca-Cola.[19] While working at the White House, Browner dealt with the Gulf oil disaster in 2010. Ironically, Browner was at the helm of EPA in 1999 when Henry Longest, acting administrator for ORD, shut down my investigations into the potential for a deep water oil rig disaster to contaminate the Gulf Stream. It was the only project at EPA or, so far as I know, anywhere else looking into the possibility of such a catastrophe and developing ways to mitigate its impact.[20]

After causing many of the nation's best scientists to leave the CDC and EPA en masse, Gerberding and Browner went on to work for the vaccine industry and other corporate giants that have a huge vested interest in maintaining an upper hand on what gets published in the scientific literature. As mentioned earlier, I asked the associate director for the environment at the White House Office of Science & Technology Policy in 1996 whether he had ever spoken with Browner about the decimation of EPA's research organization. He said he "went around and around with her over this, but was overruled at the highest level." Ironically, the defanging of science by government and industry was driven from the top floor of the Eisenhower Executive Office Building, where the president and other top administration officials work. Unfortunately, there was no Office of Scientific Integrity on the top floor to prevent the wholesale elimination of the CDC's and EPA's senior-level career scientists, many of whom who were not shy about pushing back when policymakers in Washington, DC, got too chummy with industry. The White House has effectively torn down the only firewall stopping future CDC and EPA administrations from manipulating science at will to support government policies and industry practices and placing public health and the environment at great risk.

A classic example of how things worked before then is the example set when EPA's field scientists refused to let the agency's sludge rule pass an internal peer review in 1992. Our laboratory management in Athens, Georgia, even rejected all in-house funding from EPA headquarters that was dedicated to supporting the rule. Another example is how scientists at the CDC, FDA and EPA, and the heads of these agencies under President George H. W. Bush, cooperated with me to correct problems with CDC and FDA guidelines for infection control in dentistry. Now, federal scientists working at field laboratories stand little chance of internally correcting scientifically insupportable policies handed down from Washington, DC.

In a nutshell, that is how the executive branch cleared the way for political appointees and senior career executives to use powerful corporations and leading universities to support government policies, silence top scientists, jeopardize our health, and protect corporate profits. The ultimate outcome of this transformation was that the public can no longer trust science whenever government and industry have a stake in it. Accordingly, the basis for public confidence in any court decisions that turn upon the kind of science produced by this system is also put at risk.

So, how does the public ever know for sure when to trust the scientific literature and when not to? How, for example, can the public be sure whether MMR vaccines cause autism? I'm a research microbiologist, and I don't know how to be sure myself. That area of science could not possibly be more politically charged; in such circumstances it's difficult to do honest science. Perhaps something important is being overlooked, even intentionally neglected. It's not the research that's being done for political and material gains that concerns me the most. It's the research that's not being done for all the same reasons. Sometimes a few simple experiments could change the game overnight.

It's time for the scientific community to own up to the pervasive politicization and commercialization of the scientific enterprise. And it needs to start taking a different approach when dealing with the growing chasm between what scientists publish versus what various segments of the population accept as truthful and accurate.

The public is growing more skeptical of what's published in even the most prestigious science and medical journals. Over time, public support for funding science will likely suffer, and public resistance to government mandates based on science it no longer trusts will gain momentum. The only reasonable solution is to fix the underlying problems with how science is done.

Dealing with My Adversaries

Several years ago, Stephen Kohn, director of the National Whistleblowers Center, asked me to help his son Max, a sixth-grader at Sidwell Friends School, with a school assignment. He was given the task of interviewing a scientist. Below is one of the questions he asked, along with my response.

What would you recommend to a student interested in becoming a scientist?

You need to develop the character traits of a reputable scientist. First, you have to learn to be completely honest. For example, you must reveal any data that do not support your hypotheses or conclusions, and not publish only findings that support whatever you think is true. It's also important to give others credit for what they accomplish. The way to get ahead is to help others get ahead as well.

Rather than waiting on others to point out your mistakes, be the first to discover and acknowledge them. Sometimes, that may even mean withdrawing a paper or publishing a correction when you find out later that something was not done correctly.

Learn to get along with other people even when you disagree. One day, you may discover you were wrong. Treat everyone with respect, and learn to appreciate your enemies as much as your friends; you will need them both to be the best you can be. Be quick to forgive others, and learn to forgive yourself as well.

Remember, life is short even when it's long. Picture yourself at the end looking back, and try to think of all the things you wish you had said or done. Make sure you take care of them now while you still have the chance.

Of all the advice I gave Max, perhaps the most difficult lesson for me to learn was how important my adversaries are to achieving my goals. Most people discover sooner or later that adversity can prepare them for success. But learning to appreciate those who create that adversity can be difficult, especially when their objective is to destroy their opponents' personal and professional reputations, and put them out of business.

The Maryland Court of Appeals raised some sobering questions about the medical experiments in which EPA and Johns Hopkins University exposed economically and educationally disadvantaged test subjects to paint dusts contaminated with lead. There is cause to be concerned about the methods EPA and other federal agencies are using to support their policies. But this isn't Nazi Germany, and we are not at war with government and industry or any of the academic institutions they fund. In the narrative that follows, therefore, I want to talk about the conflicts I had with EPA headquarters, the biosolids industry, and others, and how I believe these kinds of battles should be handled.

You're Hurting Us

In October 2000, my cell phone rang just as I left home to drive to UGA. It was Finis Williams, the attorney in New Hampshire who represented the Marshall family in their lawsuit against Synagro over the death of their son, Shayne Conner. EPA had approved me to serve as an expert witness for the

Marshalls in order to obtain access to medical records and other documents needed in my research at UGA.

Several weeks earlier, Synagro VP Robert O'Dette had called and invited me to speak at a session he organized at a Water Environment Federation conference in Anaheim, California. I accepted. Then Mr. O'Dette offered to fund my research at UGA. I declined. Before hanging up, he offered once again; again I declined. Apparently, the call didn't go as Synagro had planned. After turning down their offer to fund my research, Synagro's executives were worried about what I might say at the conference.

"Synagro's General Counsel, Alvin Thomas, is warning you not to say anything about Synagro in your presentation," Finis said over the phone. "He has a message for you, and he wants me to deliver it verbatim. In fact, he made me repeat it back to him over the phone to be sure that I got it exactly right, word for word."

"What is it?" I asked." "You're hurting us," Finis said. There was a long pause, then Finis said, "You know what that means, don't you?" I said, "Let me guess—it's a mafia warning." "That's exactly what it is," Finis replied.

It reminded me of something a US marshal told me when I was detailed to Florida's assistant commissioner of agriculture in 1990. Unrelated to my official duties, I helped the FBI look out for a drug dealer operating in my neighborhood in Tallahassee, where my wife and I were renting a condo for the summer. After we moved back to Athens, Georgia, I began getting phone calls at the stroke of midnight from someone with what sounded like a familiar accent. I thought I recognized it from doing some work in Columbia, SA. He would just say, "This is Carlos," then hang up.

The calls continued for a month or longer. Eventually, I called my contact at the FBI and said, "I think maybe Carlos the Jackal is sending me a warning." "Don't worry," he said, "the Jackal doesn't warn anyone." So, along the same lines, I assumed if anyone at Synagro really meant to do me any physical harm, their general counsel wouldn't be calling Finis to let him know.

At the meeting in Anaheim, the large ballroom was packed wall to wall, standing room only. Hundreds of wastewater treatment industry representatives gathered to hear what on Earth I would have to say at a session organized by my personal nemesis, Synagro. Alvin Thomas was seated in the front row directly facing the speaker's podium. In the back stood Ross Patton, Synagro's CEO. Robert O'Dette, Synagro's VP for government relations, stepped up to the podium and introduced me.

The silence was palpable as I laid a transparency on the overhead projector. It was a letter Synagro executives sent to UGA president Michael Adams, which implicated me in misusing federal and state funds. They were accusing me of criminal violations punishable by fines and imprisonment. "Silencing me is not going to solve any of the problems," I said. "We need to work together."

Synagro slide presentation obtained in discovery (with Bates numbering stamps). Courtesy of David Lewis.

Alvin Thomas exploded with allegations of criminal fraud against me, alleging that EPA had not approved my research on biosolids, and, therefore, I was misusing tax dollars. I saw O'Dette glance to the back of the room to catch Patton's reaction. Continuing nonstop, Alvin paid no attention as someone in the audience raised his voice in protest: "Let him speak!" As time passed, others began to shout in unison: "Let him speak!" It was just like the audience viewing Congress in the 1939 film *Mr. Smith Goes to Washington*, and shouting the same phrase as opponents tried to keep Jimmy Stewart's character from defending himself against false allegations of political corruption. I remember the movie because my father took my mother to see it on their first date, while he was home on leave from training for the Royal Canadian Air Force. Alvin's tirade eventually ended, but only when he could hardly be heard above the rhythmic ostinati of protests.—At which point I finished my presentation and urged once again that we all work together.

A similar spectacle played out several years later at a locally televised public hearing before the City and County Council of Honolulu. Synagro's bid to process Honolulu's biosolids had been put before the council for final approval.

O'Dette was given an opportunity to pitch Synagro's Class A biosolids, which the company claimed was pathogen free. The council asked me to follow O'Dette's presentation with a discussion of our research on biosolids at UGA.

In his presentation, O'Dette vehemently attacked my credibility. He also disputed the validity of our findings, which indicated that many disease-causing bacteria and viruses can survive Synagro's Class A treatment process, and that certain bacteria, such as *Staphylococcus aureus*, can regrow once the material is stockpiled or spread on land.

When O'Dette finished, I came to the podium and outlined some simple experiments that the council could conduct to determine whether pathogens in Synagro's biosolids could proliferate when the material is spread on land. When my talk ended, a council member alluded to O'Dette's assault on my credibility, and said, "You must really hate Synagro." I answered, "No, quite the opposite. Synagro has lot of experience and technical knowledge. Its participation is vital."

The council decided to delay approval of Synagro's contract until the tests I recommended could be done. Synagro's supporters on the council, however, were adamantly opposed to any testing. To pass the measure, the council only permitted testing for one pathogen, *Salmonella*, which is easily killed by processes used to treat biosolids. The University of Hawaii conducted the experiments, and found no *Salmonella* in the biosolids samples provided by Synagro. It did, however, discover high levels of other bacteria associated with sewage, which proliferated when the biosolids were applied to soil.[21] But, in accordance with the measure passed by the council, it did not identify the species.

The last time I saw Alvin Thomas was at a particularly contentious meeting in Atlanta. When it was over, I grabbed his arm from behind as he started toward the door. As he turned around, I said, "Alvin, when this is all over, we are still going to be friends, aren't we?" He looked me straight in the eye and said, "Yes."

The research my coauthors and I published concerning adverse health effects associated with biosolids was funded by my Department of Labor settlements. I derive a lot of satisfaction from knowing that false allegations of ethics violations and research misconduct published by Synagro and EPA headquarters gave me the wherewithal to document some of the problems with their policies and practices in the scientific literature. Perhaps in a perfect world, EPA would never have allowed Synagro and certain individuals at EPA headquarters to do what they did. But we don't live in a perfect world, and never will. That being the case, we must figure out how to make the system ultimately work for the good.

A practitioner of Hoodoo, a form of folk magic, once explained that although people will surely suffer the consequences of their actions, "God has something to do with [it]...good or bad"[22] Although I never heard it taught in church, the Book of Isaiah says of God: "I form the light, and create darkness; I make peace, and create evil; I the Lord do all these things."[23] Progress always involves a struggle between ignorance and enlightenment, good and evil, right and wrong. That doesn't mean that evil is any less evil. Sometimes, working with our adversaries can be the shortest path to the best possible outcome. The ability to do that, at least for me, comes from realizing that our adversaries can often play a vital role in making the world a better place.

When the Labor Department let EPA off the hook because it punished John Walker for distributing Synagro's allegations and took steps to prevent it from happening again, I disagreed with the decision and filed an appeal. Still, I emailed my congratulations to EPA's attorney, David Guerrero. I told him how much I appreciated how he conducted himself during the proceedings, and how highly my attorney, Stephen Kohn, spoke of the job he did.

EPA's Office of Inspector General, which a decade earlier was knocking on scientists' doors with armed agents from the Justice Department in tow, was very supportive of my efforts to change EPA policies on biosolids. It produced a report on the issues raised in my Labor cases, and concluded that EPA could not assure the public that land application of biosolids is safe.[24] And when I lost my appeal before the 11th Circuit Court in 2011, the Labor Department's attorney hurried to catch up as I left the courtroom. As she stopped me in the doorway, she said, "Dr. Lewis, I want you to know that sometimes I don't like having to do my job. This is one of those times."

Even Carol Browner went out of her way to spend time with me at an awards ceremony in Washington, DC, in 2000 when I won a Science Achievement Award for my second *Nature* article. She was very gracious, especially considering that *Reader's Digest* had just published my photo next to hers with a quote from me underneath, saying, "I have no faith in the agency or the character of its leaders."[25] I was one in a long line of employees who silently filed across the stage to receive a handshake and a smile as Browner picked up each award from a table and walked over to join her deputy for the official photo. As I started across the stage, Administrator Browner picked up my plaque from the table; while I was still some distance away, she said out loud, "I have a question for you." Her timing was such that she had to speak loudly enough that the audience would hear. More important, I would have to walk halfway across the stage wondering what in the world was about to happen.

She met me partway across the stage to speak with me face to face. The award read: *For research demonstrating that major environmental changes may substantially alter the relative persistence of the enantiomers of chiral pollutants.* She pointed to the word *enantiomers* and asked me what it meant. As I began to explain in laymen's terms, she abruptly cut me off. "No. I want you to talk with me just as you would talk with another scientist." When I was through, she said, "That's very interesting, and I think it's very important work." I congratulated her on an award she received earlier, and we continued to talk for several minutes.

Dr. Bernard Goldstein, EPA assistant administrator for ORD under President Reagan, wrote a strong letter of recommendation for my promotion to GS-15 in 2000. He alluded to the agency's "heavy-handed" approach in dealing my criticisms published in my 1996 *Nature* commentary. He also spoke of the quality of the research in my article published by *Nature* in 1999, which raised concerns about EPA's policies on biosolids:[26]

> His first authored publication in Nature *last year is an example of Dr. Lewis at his best...the kind of work which identifies Dr. Lewis as having a direct influence on environmental policy as well as an indirect role through improving the scientific stature of EPA....*

> I had an opportunity to spend some time with Dr. Lewis at a United Nations Conference I co-chaired in April 2000. He is thoughtful, articulate, well informed and cares very much about the role of science in general, and the Athens Laboratory in particular....His talk was outstanding, covering a large range of topics clearly and with authority.

Upon her retirement in 2008, my local EPA lab director provided the following public statement:[27]

Dr. Lewis' involuntary termination over his research articles was not supported by the local lab management in Athens. He was an excellent researcher and an asset to EPA science.

A Better Future Awaits

Several weeks ago, on a sunny Saturday morning, as my youngest son Jedd turned twenty-six, I unbuckled my seatbelt in one of the few B-17 bombers still flying. Bracing myself as best I could, I inched forward feet-first through the opening to a narrow tunnel under the pilots' seats leading to the plastic bubble on the nose of the plane where the bombardier sits. As I crawled toward the blinding light at the end of the tunnel, the air turbulence banged my head against the metal framework, first on one side then on the other. As I sat in the bombardier's seat, I thought about what it was like for Lieutenant Ed Knowlden, a WWII B-17 pilot I met the day before, to go on a mission. He was the only survivor on a B-17 downed in 1944 after taking FLAK over

View from a B-17 bombardier's seat. Photo courtesy of David Lewis.

Germany. At ninety-one, his memories of the prisoner of war camp outside Nuremberg are as clear as the cloudless sky that engulfed my entire view from above my head to below my feet.

That night at dinner, someone asked Ed what he thought about all the goings on in Washington, DC. "They're crazy," he replied. "That's why I don't call myself a Democrat or a Republican." The next day, he and I sat outside in rocking chairs watching planes take off and land at the regional airport in Thomasville, Georgia, where B-17s once ferried high-ranking officers in and out during the war. He talked about the day he stared out his cell window in Stalag III-A as a twin-engine British Mosquito dropped a single bomb. Missing its intended target—the train station in Nuremberg—it landed close to his building. He said the blast knocked him flat on his back, but no one in the camp was seriously injured. The next day he walked around the barracks, picking up pieces of shrapnel from the ground. I told him about Caroline Snyder's remembering walking home from school one day and British planes strafing the street, killing pedestrians. "Did Allied pilots strafe civilians, even women and children, because the Germans fired V-1 and V-2 rockets at London," I asked? "No, I don't think so," he said. "Sometimes you didn't know what you were shooting at when strafing. You couldn't tell men from women or children."

How Ed Knowlden ended up in Thomasville climbing into the co-pilot's seat of a B-17 bomber for the first time since he bailed out of one over the North Sea seventy years ago was the most amazing part of the story to me. The pilot sitting next to him on that mission, Lieutenant Joseph Kirkpatrick McClurkin, Jr., was the uncle of my best friend, Mike Woodham. Ed looked back as he parachuted through the clouds just in time to see his plane hit the water and sink. Lieutenant McClurkin stayed at the controls to keep the heavily damaged plane level, and ordered everyone else to bail out. Ed said that he saw several parachutes, but no one else survived in the frigid waters. As soon as he hit the water, a German boat crew pulled him out and treated him for hypothermia. A German officer on board, who studied at Ohio University before the United States entered the war, came in to interrogate him. Ed said the interrogator only asked one question. Alluding to a popular jazz song that came out in 1938, he asked, "What's a flat foot floogie with a floy floy?" The song was originally titled "Flat Foot Floozie," and "floy floy" was street code for venereal disease. Back in those days, you couldn't say things like "floozie" or "VD" on the radio. When Ed said that he had no idea what the words meant, the German officer replied that he didn't, either. They both laughed; and that was the end of the interrogation.

After the war, Ed Knowlden sat down and wrote lengthy letters to family members of his nine fellow airmen, including McClurkin's sister, Anne Woodham. He wanted them to know every detail he could recall about the mission that took the crewmembers' lives. Mike's mother read them to him and his sister Marianne when they were children. It had a lasting impact on Mike in particular, leading him to forget playing baseball and become a commercial airline pilot instead. Eventually, Mike located Knowlden at his home in Binghamton, New York, and went to visit him. Ed returned the favor by accepting an invitation to take a spin in the Liberty Foundation's B-17 that was stopping over at the Thomasville airport, which Mike now manages. Joining him on the flight was Mike's cousin, Joseph Kirkpatrick McClurkin III. He was born just days before his father, who captained the ill-fated mission, died ditching the B-17 in the North Sea after steadying it just long enough for his copilot Ed and the other crewmembers to bail out. While sitting in a B-17-bomber seventy years later, the man his father saved was there to talk with him face to face about how his father died.

Mike Woodham (L) and Ed Knowlden in the B-17. Photo courtesy of David Lewis.

It's important to keep looking back, remembering the source of America's greatness, and what made our nation a beacon of light shining in a world of darkness. If we ever forget how we got there, we may never find our way back when we go astray. America's role in WWII is a good place to begin, especially while the few remaining heroes of those days are fast disappearing, and the children to whom they passed on their values are leaving the work-force. More than any other war, the people who fought in WWII, like my father and Ed Knowlden, and Ed's crewmembers who lost their lives, fought to defend democracy. They risked so much, and lost so much, to protect our freedoms. As I sat in the bombardier's seat at the front of a B-17 bomber, I didn't just think about what it was like for Ed and his crewmembers to go on that fateful mission over Germany that cost all but one of them their lives, and put Ed in a German prisoner of war camp. As we flew over the area around the house where I grew up, I thought about how much I wanted to become a scientist when I explored the surrounding woods and fields as a child. I couldn't help but think about how EPA took away some of the most important freedoms Ed and his crewmembers risked their lives in WWII to protect. Seeing my world from the nose of a B-17 made me realize, for the first time, what a precious thing I had lost. Not because of anything I've experienced, but because of all that Ed and so many others in WWII suffered so that their children could grow up and be whatever they wanted to be, to pursue happiness and speak out against injustices, to not live in fear of their government. I thought about how much Lieutenant McClurkin's son, who never got to see his father, had lost. I don't think that the kinds of people who ran our government when Lieutenant McClurkin and his crewmembers went off to war in foreign lands to defend our freedoms came from the same mold as those who in recent years have begun to take those freedoms away. No one, not even the president of the United States, has the right to take any of them away.

Remembering the Holocaust, I like to imagine that the souls of those who once occupied Nazi concentration camps throughout Europe still gather in solemn assemblies where their huddled masses lived out their last days in fear and pain. And as the sun sets over vacant fields and stone memorials, I imagine them flying to the ends of the earth wherever the seeds of social injustice sprout and begin to take root. And there they preside over the living as they go about their business, reminding them of the horrors of the past and how it all began. I even like to imagine them standing ever so near, whispering the words graven beneath the feet of Mother Liberty as the Maryland Court of Appeals considered the government's experiments on poverty-stricken residents in Baltimore:

> Give me your tired, your poor,
> Your huddled masses yearning to breathe free,
> The wretched refuse of your teeming shore.
> Send these, the homeless, tempest-tossed, to me:
> I lift my lamp beside the golden door.

Perhaps never before have science and private industry collaborated in search of material gains to the extent that they do now. Consumers are becoming unwitting, and in some cases unwilling, test subjects in experiments carried out by government officials working with private industry. The products they sell may change, but their objective stays the same, which is to make a killing without anyone finding out how many people they kill. To this end, government and industry manipulate science to promote nonexistent benefits and create the illusion of safety. On the other hand, scientists who publish valid data in the public interest are increasingly at risk of having their careers ended and reputations destroyed. And, as we see with those who are drawn to nontraditional medicine, or question vaccine safety, people who criticize their methods or raise concerns about their products are portrayed as ignorant, uneducated, or even a threat to the public welfare.[28]

12

CHANGING DIRECTION

B y republishing Synagro Corporation's allegations of research misconduct, Brian Deer and the biosolids industry are fulfilling a promise made by Synagro's vice president for government relations, Robert O'Dette. After O'Dette published the allegations in 2001, he wrote, "These people will find their work rightly discredited and their funding will disappear while credible researchers continue to have funding."[1] "Credible" researchers, of course, are individuals paid by Synagro and others to publish research portraying their commercial products as safe and beneficial. Below are some of the projects I planned at EPA, which I hoped would alter the direction of science in several important areas. I still hope to pursue some of this research myself one day.

Environmental Health

The overarching objective of pollution control should be to reduce complex mixtures of industrial pollutants in air, water, and soil. From its inception, EPA has devoted its resources to removing a few trace chemical pollutants from air and water while promoting land application of biosolids, which contain a whole universe of fat-soluble municipal and industrial pollutants at orders of magnitude higher concentrations. While the FDA takes into consideration drug interactions, EPA acknowledges the importance of synergistic effects, but has not integrated this reality into its approach to regulating chemical pollutants.[2] As discussed in chapter 3, the chances that any living organism could suffer irreparable harm from exposure to traces of a single pollutant is most likely very tiny when compared with exposures to highly complex mixtures. Serious adverse effects resulting from exposure to sufficiently high (ppb and higher) concentrations of thousands of chemical pollutants in complex mixtures is a virtual certainty, especially during early developmental stages and over multiple generations. Biosolids are a worst-case scenario. EPA should require that all sewage sludges collecting at wastewater treatment plants be destroyed using extreme heat, perhaps as a component of waste-to-energy programs.[3]

Research I planned at North Carolina State University in Raleigh, North Carolina, combined my background in exposures to complex mixtures of environmental pollutants with studies of pollutant-related DNA damage pioneered by Richard Winn. Using large populations of transgenic fish (medaka) to detect very low mutation rates, Dr. Winn discovered that persistent DNA damage causes new mutations to occur during prenatal development.[4] Because these mutations occur in early-stage embryos, they can potentially lead to any of a number of developmental problems, including neurological disorders such as those associated with autism.

Most interestingly, he discovered evidence that error-prone DNA repair processes can create new mutations that never existed in either parent. Traditional methods used to investigate any genetic components that may predispose individuals to environmentally triggered diseases and disorders don't look for this type of indirect mutagenesis. Our research, therefore, would combine two very important aspects of research on environmentally triggered diseases that are currently being overlooked. It could open up a new front in the search for causes of autism, Parkinson's disease, multiple sclerosis, rheumatoid arthritis, Crohn's disease, and other neurological and immunological diseases and disorders. Even a CDC scientist, who vehemently objected to our including vaccine adjuvants in our research, conceded:[5]

> *I have no problem and in fact would support two of Dr. Lewis'*
> *aims, i.e., "Broaden the range of autism-related environmental*
> *and genetic factors studied by environmental health scientists*
> *to include...complex mixtures of industrial wastes [and] muta-*
> *tions that occur in developing embryos as a result of the parents'*
> *exposures to complex mixtures of environmental pollutants....*
> *These are worthwhile aims. In particular, I and others have*
> *been pushing for at least 15 years for more research to deter-*
> *mine whether in utero exposures to complex mixtures of envi-*
> *ronmental pollutants is a cause of autism.*

Antibiotic Resistance

One of the biggest problems facing the practice of medicine is deadly pathogens such as carbapenem-resistant *Enterobacteriaceae* (CREs) and methicillin-resistant *Staphylococcus aureus* (MRSA) rapidly becoming resistant to even our last line of defense antibiotics. Antibiotic resistance genes (ARGs) spread from one species of bacteria to another by sharing strands of DNA called plasmids. Physicians over-prescribing antibiotics, patients not following directions, and concentrated animal-feeding operations (CAFOs) are thought to be the most common sources of the problem. Information on some of the most important sources, however, is being suppressed by EPA, USDA, and perhaps other government agencies as well. For example, a research microbiologist who left the USDA reported that he was prevented on at least eleven occasions from publicizing his research on antibiotic-resistant airborne bacteria originating in farm wastes.[6]

My coauthors, which included a pediatrician treating patients exposed to biosolids, and I found that about one-fourth of the residents at ten land-application sites suffered from recurring *Staph aureus* infections. Many were resistant to one or more of our most effective antibiotics. As discussed in previous chapters, the head of EPA's Office of Wastewater Management met with Synagro executives in 2001 to discuss how to discredit our research, and, to that end, one of the EPA scientists who attended the meetings publicly distributed Synagro's false allegations of research misconduct against me. Since then, others have discovered that ARGs at wastewater treatment plants spread from biosolids, biofilms, and sediments to receiving rivers.[7] Most antibiotics are consumed by roughly ten thousand CAFOs in the United States.[8] Wastewater treatment plants, by contrast, are located in every city and town in the United States. These plants receive wastes from most people living in the country, as well as from hospitals and other facilities using antibiotics of last resort. Most of the sewage sludge collected at wastewater treatment plants is spread on public and private lands including farms, parks, golf courses, school playgrounds, and home gardens. EPA's Targeted National Sewage Sludge Survey (TNSSS) of seventy-four wastewater treatment plants across the United States in 2009 found high (ppm) levels of antibiotics in sewage sludges.[9] Biosolids are a major reservoir of ARGs; hence, biosolids operations spread ARGs along with all antibiotics in use to soils, which, in turn, transport them in dusts and water runoff throughout our environment. Biosolids operations, therefore, potentially account for much, if not most, of the problems associated with antibiotic resistance. At the same time it is, by far, also the least studied source, and public health information regarding biosolids-related problems is subject to the most intense government suppression.

Honey Bee Colony Collapse Disorder

Since land application of sewage began in earnest in the 1980s, commercial honeybee operations in the United States and elsewhere throughout the world have reported losing massive numbers of worker bees. In commercial agriculture alone, approximately one-third of the world's food crops depend on honeybees for pollination. Because of an inexplicable phenomenon called "colony collapse disorder" (CCD), functionally impaired worker bees are abandoning their colonies in irreplaceable numbers. So far, CCD has resulted in a loss of 50 to 90 percent of colonies in beekeeping operations across the United States and we are now importing bee colonies from Australia to maintain our current levels of agricultural production.[10] Some experts estimate that honeybee populations in the United States will be completely wiped out by 2035, in which case all colonies used in agriculture will have to be imported from abroad each growing season. With other parts of the world experiencing the same problem, there is no guarantee that honeybee populations anywhere else will remain healthy.

The main symptoms, neurological effects and suppression of the immune system, are attributed to complex mixtures of pesticides and other agricultural chemicals, which are found in bee pollen at high-ppm levels.[11] In 2013, the European Union placed a temporary (two-year) ban on highly persistent neonicotinoid pesticides, which are in widespread use and can produce the kinds of neurological effects observed. Even scientists at EPA headquarters expressed concerns over studies the agency used to approve these chemicals.

Before my visiting scientist position at UGA ended in 2008, I wanted to compare CCD rates among honeybee colonies pollinating the same crops on farms treated with and without biosolids that contained high inputs of wide varieties of mixtures of industrial wastes, which include pesticides and a plethora of other neurotoxins. This work would have been similar to previous research combining microbiology and entomology, which I conducted at EPA in collaboration with the USDA.[12] In those studies, my coworkers and I looked at the metabolism of pesticides and other chemicals by micro-organisms in the digestive tracts of insects.

Vaccine Safety

The CDC and other government organizations assure the public that vaccines do not cause autism and that risks of adverse reactions of any kind are extremely small. As with biosolids, however, the government's reassurances are based on a body of studies funded by government agencies and large corporations with vested interests in downplaying problems with vaccines. And, as with researchers who associate biosolids with adverse health effects, scientists who link vaccines to adverse health effects risk losing their jobs and reputations. Whenever these conditions exist, the public has good reason to be concerned.

I recently asked a large animal veterinarian in my area whether she was seeing much in the way of adverse reactions when vaccinating animals. She said, "It used to be a problem until several years ago." "What changed?" I said. "We learned to spread them out," she replied. Again, I don't want to keep beating a dead horse, but vaccine science and biosolids science seem to have a lot in common.

For one thing, government regulators responsible for both biosolids and vaccines ignore the potential for chemical ingredient interactions. Whether a company manufactures a new chemical that pollutes the environment or a new vaccine, it isn't required to test its effects when mixed with other chemicals. With vaccines, for example, manufacturers are not required to test the effects of vaccines when administered with adjuvants and other chemical and biological agents contained in the final vaccine preparations.

Every year, we see cases where young, otherwise healthy individuals suddenly die from mixing four or five prescription drugs. If everyone took these same combinations, severe illness and death would commonly occur. So why should anyone assume that similar reactions don't occur when thousands of chemicals, including many of these same pharmaceuticals, are mixed together in biosolids, often in high ppm concentrations? Along these same lines, the government needs to develop a more appropriate response to public concerns over children taking increasing numbers of vaccines at the same time. At a minimum, the CDC should recommend mandatory reporting of any severe adverse reactions that patients and caretakers associate with vaccines. This should include any effects that may not be evident for weeks or longer.

Space Exploration

When I moved to Athens, Georgia, to go to college, I spent time with a Greek couple working in a local restaurant. The owner, who was also Greek, was a wealthy man with a reputation of being ill-tempered. He took a liking to me for some reason, but didn't seem to care much for people in general. One day he returned from Greece a changed man, and was never the same again. Overnight, he became overly compassionate toward everyone, even complete strangers. It was a long time before I finally asked him what brought about this remarkable change in his disposition. He said he just looked out his window as his plane approached the ancient city of Athens where he grew up. "All the people looked like little ants. For the first time in my life," he said, "I realized how small and unimportant I am."

By proving that the sun, and not the earth, is the center of our solar system, Galileo, Copernicus, and others transformed both science and religion. Progress in science and religion go hand in hand, therefore, expanding the heart as well as the mind. True humility is a precious gift. For me, it took the Hubble telescope to make it really sink in how utterly insignificant I really am. The things we will see as we explore the solar system and beyond will continue to change how we view ourselves. Once we see and experience extraterrestrial life as it exists beyond the boundaries of Earth, humanity will begin to enjoy a completely different level of understanding and compassion for all forms of life on Earth.

Find Life, Spread Life

During the space race in the 1960s, I built my own rockets and launched them in a remote field. Recently, I heard a NASA engineer reflect upon his early days of experimenting with homemade rockets, which he aptly called "pipe bombs with fins." My best friend Mike Woodham and I narrowly escaped injury by hitting the ground as one malfunctioning missile passed several feet overhead and impacted the trunk of a large oak tree we were standing under. At night, I kept track of Soviet and American satellites and estimated their altitudes based on their positions in the night sky when they stopped reflecting sunlight. The Federal Aviation Administration and Moody Air Force Base in Valdosta, Georiga, accommodated some experiments Mike and I did for a junior high school science project. In a magazine article, "They Flew to 65,000 Feet...The Easy Way," an Air Force reporter wrote, "Through the combined efforts of certain segments of two large and very busy federal agencies, two small boys, whose scientific knowledge was furthered, may some day make their own great contributions to the Space Age." It's ironic looking back, because two large federal agencies, EPA and USDA, put an end to my research career.

During my last years at EPA, I was fortunate enough to visit NASA's Ames Research Laboratory in California, and begin collaborating with their scientists to develop closed biological systems for supporting astronauts on long-term missions to Mars and elsewhere. My feeling about NASA's manned space program, however, is that budgetary constraints are likely to preclude such missions for the foreseeable future. Personally, I would like to see NASA's priorities take a different direction. Searching for life elsewhere in the solar system, and spreading complex life on Earth to our neighboring planets and beyond, should be its highest basic research priorities.

One of humanity's greatest responsibilities, I believe, is to give Earth's complex life forms a chance to spread throughout the universe. We have had these capabilities for several decades now. To assume we will have them for centuries or millennia to come is presumptuous. Looking back in geological time, it's pretty clear that the earth couldn't have supported billions of people during much of its past. Our window of opportunity for finding and spreading life elsewhere in the universe could suddenly come to a close during any given generation, including the present one. We should act now, first by systematically searching for any evidence of life where solid or liquid water exists, or has existed, in our solar system. As soon as we can reasonably rule out the probability that it exists in any of these places, we should seed them with our own forms of life. Then we should push on beyond the boundaries of our solar system as quickly as possible. To me, it makes more sense to send humans to Mars and elsewhere after pockets of life have been established. That way they will have some means of prolonging their survival if problems arise with returning to Earth.

Our oldest son, Josh, is working on his Ph.D. in cell biology at Emory University in Atlanta. Lately, we've been talking about how some life forms, such as tardigrades, may survive in other worlds, and how microorganisms could be engineered to survive and rapidly evolve into plants and animals. Tardigrades are tiny animals that feed on plants, algae, and small invertebrates. They can survive even the hard vacuum of deep space and crushing pressures greater than in the deepest ocean trenches by undergoing a reversible state of suspended animation. Although genetically modified organisms (GMOs) are not welcome among environmental activists, perhaps they can find happiness in other worlds. Alien life forms somewhere in the universe may even share their manmade DNA sequences.

Here are some of Josh's ideas:[14]

> *Many strategies are possible for spreading complex life to other worlds. For example, cyanobacteria (blue-green algae) could be used to create an oxygen-rich atmosphere on planets with high concentrations of atmospheric CO_2, just as they did on Earth billions of years ago. To do the same on other planets, they would probably have to be genetically engineered to survive alien conditions. To speed up evolution, it wouldn't be difficult to insert dozens of "advanced" genes—genes commonly used by higher organisms to create multi-cellular structures—into the cyanobacteria's genomes. The basic machinery for producing adhesive cellular junctions, intercellular signaling, organism-level development, and patterning could all be stashed away for future use when oxygen levels rise.*

> *To ensure their survival, however, these highly advanced genes would be have to be repressed until oxygen levels are suitable for more complex forms of life. Otherwise, they will be dumped as excess genetic baggage. Nature has various strategies for maintaining genes that become important at some point in the future when conditions change as they have in the past. One effective strategy nature employs is to place "toxin–antitoxin cassettes" immediately adjacent to important, but rarely used, genes. These cassettes are comprised of stretches of DNA encoded to produce toxins side-by-side with stretches of DNA encoded to produce their antidotes, called antitoxins. Usually, the toxin is stable, while the antitoxin is not.*

> *If a mutation happens to eliminate the temporarily "useless" gene, the toxin–antitoxin cassette is lost as well. Instantly, the production of both toxins and antitoxins halts. Since the antitoxins are chemically unstable, they quickly break down. Then the*

remaining toxins kill all of the mutant cells missing the "useless"
genes. This same strategy could be engineered into cyanobacteria
to preserve genes they will need for evolving into plants and
animals as atmospheric oxygen levels rise.

Over time, mutations occurring during reproduction would
eventually turn on some, if not most, of the advanced genes for
producing complex structures. Then selective pressures should
steer the organisms' evolutionary pathways toward complex,
multi-cellular life very similar to what we have on Earth.

Thus, by building such systems into microorganisms used to
seed other worlds, the evolutionary process that took billions of
years to go from bacteria to highly complex multi-cellular organ-
isms on Earth could be engineered to occur within much shorter
spans of time on other planets.

It would be incredibly interesting to artificially seed lifeless planets and
moons with some of the earth's simplest life forms and watch as complex
communities develop. It would no doubt yield many insights into the different pathways life on Earth could have taken. But the only one way we will
ever fully understand life on Earth is to find where it already exists, or has
existed, in other worlds.

Cosmic Evolution of Life

Astrobiologists, scientists who search for evidence of extraterrestrial life, generally think in terms of life being generated ab initio throughout the universe
wherever conditions are conducive for living microbes to form from complex
organic molecules. Such conditions could have existed as soon as the earliest planets revolved around aging stars almost fourteen billion years ago.
Currently, scientists are also focusing on comets and other extraterrestrial
carbonaceous debris as carriers of microbial life in which, according to the
theory of *cometary panspermia*, planets are showered with viable microbes.[15]
Since a German paleontologist, Hans D. Pflug, first reported what appear to
be fossilized microbes in carbonaceous meteorites over three decades ago, a
striking array of similar examples have been discovered.[16]

Single-celled microorganisms appeared on Earth at least 3.5 billion years ago. One of the earliest and most abundant forms is a group called archaea, which lack a nucleus but possess genes and metabolic pathways found in more complex organisms. Many species of archaea can thrive under extreme conditions of temperature, pressure, and pH. They use a wide variety of energy sources, including metal ions and gases such as ammonia, methane, and hydrogen. Some even gather energy from sunlight and convert carbon dioxide to sugars as plants do. To me, this suggests that microorganisms of extraterrestrial origins may have come from planets similar to our own, where highly advanced forms of life existed. If true, extraterrestrial microbes may be embedded with fragments of advanced life forms that could allow them to steer evolution in the directions it has taken on Earth, and on other planets they may have found suitable for life. This would mean that the complexity of life we now enjoy reflects an evolutionary history that predates the age of the earth, and may trace back almost four times that timespan. In other words, the framework for many of the evolutionary leaps that life took on Earth over the past several billion years may have developed long before our solar system even existed. That means highly complex forms of life very similar to our own may commonly exist wherever suitable conditions have lasted as long they have on Earth.

In 2009, Australian microbiologists reported microbial life in caverns dating back 2.75 billion years, which "represent a new habitable microenvironment for early life on Earth, and an analogue for ancient life on Mars."[17] Personally, I would be most interested in investigating the earth's early impact by a Mars-sized planetoid, after which our moon formed from the accretion of collision debris orbiting the earth. Showering the young earth with alien microbes entrapped by the accretion process may have been just what it took to effectively inoculate the earth with life. If so, a record of that event may still exist buried beneath the lunar surface.[18]

Global Warming

My last major responsibility at EPA was conducting research on global warming. Part of my job was overseeing an EPA project headed by Dr. Jerry Melillo, Co-Director of the Ecosystems Center in Woods Hole, MA. Dr. Melillo, at the time, was Associate Director for the Environment at the White House Office of Science & Technology Policy under President Clinton. As such, he worked closely with Vice President Al Gore regarding climate change.

As with all big science issues, it's important to understand the big picture. Unlike the political debates over what's driving the current warming trend, or whether it even exists, the basic history of planet Earth not subject to much debate. The earth began as a molten planet when the solar system formed, and, in time, cooled enough for its surface to solidify. As it continued to cool, our atmosphere and oceans formed and life took hold. Although this process continues, the decay of uranium and other radioactive elements generates enough heat to significantly impede the cooling rate of the earth's core.

Eventually, the earth's core cooled enough to allow periodic glacial activity. Approximately every 120,000 years, ice builds up in polar regions and advances toward the equator. Then it recedes as it did about twelve thousand years ago. The onset of an Ice Age, or lesser cooling events sometimes called Little Ice Ages, is thought to be triggered as fresh water from glacial melt disrupt the salinity gradients that keep ocean currents moving heat away from the equator. The Gulf Stream, for example, appears to have weakened during the last Little Ice Age.[19]

In time, the core of the earth will finally solidify. The core of Mars, which is smaller than the earth, has already done so. Once that happens, the earth, like Mars, will become a frozen planet. As its solidified core no longer generates a magnetic field, solar winds will strip away most of the earth's atmosphere. In the big picture, therefore, the surface temperature of earth-like planets has a lot to do with the amount of heat emanating from their cores.

Over the past several decades, scientists have learned a lot about the warming effects of CO_2 and other greenhouse gases. This information is critically important to understanding the current warming event, and what's driving it. It may well be that fossil fuels are almost entirely responsible for the current trend toward higher ocean and atmospheric temperatures. Still, I would like to know more about what's happening inside the earth, and what role fluctuations in heat transfer from the core to Earth's surface via magma convection may have played in warming events in the past. Since the temperature differences between magma and seawater is so extreme, even the slightest shifts in magma convection on a global scale could potentially contribute to global warming by heating the oceans. Unfortunately, this area of science is not well developed.

It's true that the current rise in ocean temperatures, and the rapid melting of glaciers, are tracking closely with surface temperatures and atmospheric CO_2 concentrations from paleoclimatic evidence.[20] A simple experiment, however, can demonstrate the difficulty with establishing cause and effect.

Just turn up the heater in an aquarium and measure CO_2 concentrations in the air above it as water temperatures rise. The results will likely mirror the paleoclimatic record because, as temperature rises, so does natural CO_2 production from bacterial respiration and other biological sources.

I'm not saying that the prevailing view of what's causing climate change is wrong. I just think it's still important to get a grasp on what's happening to the earth's temperature in the larger picture, which—at least over very long time scales—is mainly driven by heat transferring from the earth's core to its surface. Of course, that will change as our sun begins to run out of hydrogen, collapses, and explodes into a Red Giant.

The reason this matters is twofold. If a substantial amount of the extra heat driving the current warming period is coming from changes inside the planet, efforts to reduce greenhouse gases won't have as much of an effect. Perhaps more importantly, if we experience another Little Ice Age in the next century or two, elevated levels of greenhouse gases could help mitigate its impact. Given the consequences of getting it wrong, we need to have a good grasp on the dynamics of both solar-related heat and heat-transfer from the earth's core to deep ocean water. Unfortunately, work in the latter area deals mainly with localized effects of magma injection, rather than heat transfer on a global scale.[21]

In closing, Secretary Kerry's remarks about the most recent findings of the Intergovernmental Panel on Climate Change, which go to the very heart of what this book is about, are very troubling.[22] He begins: "Read this report and you can't deny the reality: Unless we act dramatically and quickly, science tells us our climate and our way of life are literally in jeopardy. Denial of the science is malpractice." In a related speech, Kerry indicates that the United States will handle climate change as it does terrorism.[23] He concluded: "The bottom line is this: [the way we think about terrorism] is the same thing with climate change. And in a sense, climate change can now be considered another weapon of mass destruction, perhaps the world's most fearsome weapon of mass destruction."

To understand my concerns, I refer to my comments about the "terrors of science" that result whenever governments and the universities and other institutions they support seek to silence scientists who have genuine concerns.[24] This is especially true when it relates to a body of science that's largely controlled by federal funding. Any efforts by the U.S. Government to silence scientists who question its position on climate change will only undermine its efforts to build public support. Moreover, Kerry's remark, "[President] Obama and I believe very deeply that we do not have time for a meeting anywhere of the Flat Earth Society," sends the wrong message.[25] To strengthen public confidence in the Administration's position, its message should be something like: "Although we're convinced we are taking the right course of action, the President and I are willing to meet with credible scientists who believe we have overlooked important considerations."

EPILOGUE

When *Nature* commented on the National Academy of Sciences' (NAS,) handling of my research on biosolids, it aptly described the past three decades of EPA's biosolids program as "the failure of presidential administrations of both parties."[1] Since then President Obama has strengthened the biosolids industry's grip on EPA by reappointing Robert Perciasepe and elevating him to deputy EPA administrator. As head of EPA's Office of Water under President Clinton, Perciasepe created the Biosolids Incident Response Team (BIRT) to stop dairy farmers in Georgia and my research team at UGA from documenting illnesses and deaths linked to biosolids.[2] The NAS used BIRT's reports, which contained false and fabricated data (see Appendix V), to support its conclusion that there is no documented evidence that EPA's 503 sludge rule has failed to protect public health.[3] Perciasepe also led the attack on scientists at Cornell University who published *The Case for Caution: Recommendations for Land Application of Sewage Sludges and an Appraisal of the US EPA's Part 503 Sludge Rules*.[4]

For his second term, President Obama nominated Thomas Burke, dean of Johns Hopkins School of Public Health, to head EPA's Office of Research & Development (ORD). Burke, who chaired the NAS committee, permitted Synagro's white paper falsely accusing me of research misconduct to be considered in the panel's deliberations, and removed references in draft versions of the NAS report to my research documenting adverse health effects linked to biosolids peer-reviewed scientific literature.[5] These erasures further cleared the way for the committee to give EPA's biosolids regulation a clean bill of health. In a unexpected turn of events, however, EPA recently approved a new sewage sludge gasification technology as an economically sustainable, clean alternative to burial, land application, and incineration of biosolids. This is the kind of approach Caroline Snyder and I have recommended in this book.[6]

Acting ORD administrator Henry Longest, the architect of EPA's bio-solids programs, who made a career out of ending mine, received multiple presidential awards in Rose Garden ceremonies during the Clinton and Bush administrations.[7] Similarly, President Bush appointed Gale Buchanan, dean of UGA's School of College of Agricultural and Environmental Sciences, as under secretary of agriculture for research, education, and economics.[8] At UGA, Buchanan prevented faculty members from testifying about the fraudulent data EPA and UGA published to support EPA's 503 sludge rule, and discouraged efforts by the Department of Marine Sciences to give me a tenured faculty position."[9] Perhaps it's only coincidental that key individuals involved in covering up adverse health effects of biosolids on public health keep getting rewarded with presidential appointments and awards. But I seriously doubt it. More likely, the examples I've given based on my own experiences just reflect the historical role that the president of the United States has always played in maintaining the machinery Uncle Sam uses to create the illusion of science needed to support government policies and industry practices. And it's not just "sludge magic." Uncle Sam's magic science machine likely impacts any area of science where government policymakers have a stake in what the peer-reviewed scientific literature has to say, whether it's about food products, green energy, or anything else affecting government regulations and industry profits. In areas where the stakes are very high, such as biosolids and vaccine safety, the government's approach appears to be no-holds-barred. False allegations of research misconduct, data fabrication, and other forms of scientific misconduct are ignored, if not outright supported, at the highest levels of many of our most prestigious scientific institutions. The negative impact on the overall integrity of the scientific enterprise can no longer be overlooked.

If the White House were to ever get out of the business of manipulating science with federal funding, public trust in the scientific enterprise would increase, and people could make better choices about protecting their health and taking care of the environment. I think government and industry could thrive in such a world. The current system, based on selectively funding scientists who support the interests of government policymakers and silencing those who don't, is unsustainable. Once the public no longer trusts the scientific literature, what scientists publish will no longer matter.

In 2011, I began working with the University of British Columbia's (UBC's) Neural Dynamics Research Group to collaborate with a group of scientists in Raleigh, North Carolina. Our objective is to investigate whether exposures to complex mixtures of environmental pollutants and aluminum adjuvants in vaccines, especially during pre- and early post-natal development, could cause neurological disorders such as autism.[10]

Then, in 2012, the biosolids industry joined forces with Brian Deer, and republished Synagro's false allegations of research misconduct. In an article titled "David Lewis, Ph.D. Shifts Focus," the New Hampshire based North East Biosolids and Residuals Association (NEBRA) supported Deer's use of Synagro's allegations, and linked to Deer's website, where Deer republished Synagro's allegations.[11] NEBRA, which is sponsored by Synagro, first published the allegations in 2001.[12]

NEBRA's mission is to "share information widely, both in and around New England and eastern Canada, and through contacts around the continent and the world."[13] Its executive director, Ned Beecher, is one of three national liaisons for staffing the Water Environment Federation's (WEF's) biosolids efforts.[14] After spending ten years dealing with Synagro's false allegations, I finally had an opportunity to pick up where I left off at EPA and UGA and continue my research in Raleigh. But, after learning of the attacks on me by Deer and NEBRA, a key scientist withdrew from the project, and the effort collapsed.

During my last eight years at EPA, I published groundbreaking studies in *Nature, Lancet,* and *Nature Medicine.* My overarching goal was to develop a fundamental understanding of how quiescent (dormant) microorganisms remaining viable in soils, sediments, glaciers, deep ocean currents, and other environments, sometimes for thousands of years or longer, interact with microbes that are actively growing in the environment. I remain convinced that understanding how microbiological processes on Earth reach and maintain steady-state conditions over long periods of time rests in understanding the dynamics that go on between those two worlds—dormancy and active growth. Understanding that is vital to our survival. I believe it ultimately controls the composition of the major gases comprising our atmosphere, and important nutrient cycles upon which plant and animal life depend. It was an area I pioneered, and presented at Gordon Conferences in New England, including one full session on the topic.

When government and industry target scientists, the process isn't random. It tends to target those who have the greatest impact on the scientific community, and public opinion. They're made examples just because they are so visible. In other words, they want to send the strongest signal possible to anyone else who may be contemplating publishing information that may threaten certain government policies or industry practices.

From the beginning, I knew I would lose my job if I published papers documenting problems with EPA's biosolids program. For me, it was just a question of how important the issue was to public health, and whether I stood a reasonable chance of eventually changing the agency's direction. It was a lot to sacrifice. Having my own lab and being free to pursue my own projects was all I ever dreamed of doing since I was a child. But the life I ended up with instead was just as challenging, exciting, and rewarding.

Realizing my childhood dream of having my own research lab, if even for just a short time at EPA, made me fully appreciate the value of what I'm striving for others to experience. Scientific research is all about discovering firsthand how nature works, and using that knowledge to preserve, protect, and advance life on Earth and beyond. If forfeiting my career in research opens the door for other scientists to pursue their dreams free from unwarranted government and industry control, my sacrifice was no sacrifice at all. I'll gladly revel instead in their accomplishments, which will no doubt far surpass anything I could have ever done.

—David L. Lewis
January 8, 2014

PRESIDENT JOHN F. KENNEDY'S LEGACY

On January 20, 1961, John F. Kennedy joined President Eisenhower on Capitol Hill and took the Oath of Office. In his inaugural address, he immediately focused upon foes of the United States who seek to "alter that uncertain balance of terror that stays the hand of mankind's final war." In place of a nuclear arms race, President Kennedy admonished, "Let both sides seek to invoke the wonders of science instead of its terrors. Together let us explore the stars, conquer the deserts, eradicate disease, tap the ocean depths, and encourage the arts and commerce."

President Kennedy made his own unique contribution to science by vowing to land a man on the moon before the end of the decade, and bring him back safely. His vision inspired countless young men and women to become scientists and engineers, and the pace of scientific and technological advancement accelerated worldwide. During the half-century that followed, our fundamental understanding of the universe has been revolutionized. President Kennedy's impact on science, therefore, reached far beyond *one small step* taken by Neil Armstrong, and truly became *one giant leap* for all of humanity.

When speaking at the John F. Kennedy School of Government in 2011, I talked about how the collaborative manipulation of scientific research by government, industry, and academia is increasingly deflecting the overall direction of science toward supporting government policies and industry practices. Sheila Jasanoff, director of the school's Science Technology and Society Program, asked when I was done, "So, how do we solve this problem?" I replied, "All it takes is for the right people in the right position to take action." Then I turned to my left, and said, "Here sits Harvard University's vice provost for research, who pointed out himself that most research misconduct arises from the relationship research scientists have with their employers. Just think," I said, "what a difference it could make if someone in his position were to take on this problem, and say, 'It won't be tolerated at Harvard.'"

A system that selectively funds scientists who support government policies and industry practices while weeding out those who don't may not destroy the world as quickly as nuclear war would. But it does share the potential for harming public health and the environment on a global scale. Vaccines, pharmaceuticals, fertilizers, and a host of other commercial products that drive the world economy can save many millions of lives—or just as easily be turned into weapons of mass destruction. And, like nuclear war, the unleashing of mass destruction by science gone awry at the world's institutions of scientific research is probably more likely to be inadvertent than intentional.

President Kennedy's Death

It was not happenstance that President Kennedy's personal secretary, Evelyn Lincoln, called me the morning after my research on HIV transmission in dentistry aired on ABC's *Primetime Live* in 1992. The network's main draw for the program was an interview with JFK, Jr., which immediately followed my segment. Ms. Lincoln's assurances that President Kennedy would have personally handled the issues I raised had they come up during his administration seemed very plausible to me. I could imagine getting that same call in 1962, and Ms. Lincoln saying, "Please hold for the president." But it's hard to imagine that happening with any other presidential administration since.

When President Kennedy was assassinated, the government changed in ways no one could foresee. Key evidence that could explain much of what happened that tragic day disappeared, and conspiracy theories involving the FBI, the CIA, and even the White House abounded. These suspicions still loom heavy over Washington, DC, and the government has never fully regained the public trust. If anything, public distrust of the FBI, the CIA, and other federal agencies has only deepened with time. The government, in turn, has grown intolerant of those who question its policies. It's almost as if government integrity sat in the path of the assassin's bullets, and died with President Kennedy.

After the president's funeral, former First Lady Jacqueline Kennedy spent six weeks in seclusion on Greenwood Plantation in Thomasville,

Georgia, and attended a local Catholic church. The church, now All Saints Episcopal Church, has since been moved to one of the city's historic districts. My brother Dudley, who owns a construction company that handles projects for some of the area's plantations, took over routine maintenance of the building. To many, it is a constant reminder of the great loss that our country suffered. President Eisenhower and his wife, Mamie, also visited Greenwood Plantation, which was owned by John Hay Whitney, President Eisenhower's ambassador to Great Britain.

The Terrors of Science

The "terrors of science" associated with nuclear war, of which President Kennedy spoke in his inaugural address, bring to mind horrific images of Japanese women with charred flesh hanging from their bodies as they wandered through the rubble of Nagasaki and Hiroshima carrying their severely burned children in their arms. For others, it may trigger similar memories of the Holocaust. Seeing children with limbs deformed by thalidomide, or covered with boils from being exposed to municipal and industrial wastes in biosolids, may be less disturbing. And targeting scientists with false allegations of research misconduct, or forcing them to leave the CDC and EPA en masse, are not comparable with Nazi atrocities. But these too are the terrors of science. Left unchecked, they could eventually lead to more suffering and death than what millions experienced in WWII. The US government, therefore, bears a responsibility to ensure that it, the academic institutions it funds, and the industries it regulates also invoke the wonders of science instead of its terrors.

ABOUT THE AUTHOR

DAVID L. LEWIS, PhD, was a research microbiologist with EPA's Office of Research & Development for thirty-two years, and a visiting scientist and member of the graduate faculty at the University of Georgia. He was awarded EPA's Science Achievement Award for his groundbreaking research published in *Nature* concerning the effects of major environmental changes on the persistence of environmental pollutants. His earlier investigations published in *Lancet* and *Nature Medicine* concerning an outbreak of HIV in a Florida dental practice prompted the CDC and other public health organizations worldwide to adopt the current heat-sterilization standard for dentistry. His research on public health problems associated with treated sewage sludges (biosolids) prompted the House Science Committee to hold two hearings and Congress to pass the No Fear Act of 2002. He is currently collaborating with the Neural Dynamics Research Group at the University of British Columbia, and oversees the Research Misconduct Project at the National Whistleblowers Center (www.researchmisconduct.org). He and his wife, Kathy, who teaches music, have two children, Josh and Jedd, who are preparing to become research biologists.

AFTERWORD

I n this book, readers learned the fascinating and frightening story of the lengths to which government will go to protect its policies, even when they endanger human health, agriculture, and the environment. The unethical and illegal methods that EPA used to silence one of its own who refused to accept such a policy were described and documented in great detail.

The government's methods included deception, fraud, data manipulation, corruption of the peer review process, false accusations, and false and misleading information posted on government webpages and then disseminated to the media, as huge amounts of public funds were spent in an attempt to prevent Lewis from uncovering the body of deceptive scientific literature it created to support its policy.

Every avenue Lewis took to restore scientific integrity within his agency was met with one insurmountable roadblock after another. He persisted, filing lawsuits on behalf of injured sludge victims and farmers, arranging congressional hearings, taking the battle to influential congressional leaders of both parties, and even to the highest level at the White House.

This book exposes a completely new level of government corruption. Agencies tolerate, and even support, bad policies and yield to industry pressure to avoid strengthening regulations that put the public at risk. In this case the government itself initiated just such a policy, and then aggressively used the nation's resources to engage academia, industry-paid scientists, and industry lobbyists to help promote and protect that policy.

Then Lewis spent over a year sifting through all of the documents used in the legal proceedings against Andrew Wakefield, including key evidence that was withheld from the public, which would have exonerated this scientist. But it was too late. By then, every major science and global news organization had published Deer's allegations that Wakefield had committed scientific fraud by deliberately misrepresenting data.

Until *Nature* reporter Eugenie Reich attended Lewis's presentation at Harvard University and interviewed Fiona Godlee of the *British Medical*

Journal about the documents Lewis uncovered, no mainstream investigative reporter ever questioned the veracity of Deer's false allegations. Even if reporters had investigated the charges, the skills needed to uncover the deception could only have been provided by an expert, trained and experienced in interpreting the relevant medical data. Lewis was that expert.

Others have offered suggestions on how to deal with corporate corruption of science, and the kinds of legislative changes needed to support and protect researchers who work for the public interest.[1] It remains to be seen if any of these recommendations will be implemented to restore research integrity. Scientists working for the public interest should not be forced to devote a large proportion of their productive years to filing expensive lawsuits to clear their name and set the record straight.[2] Nor should they have to "give and hazard all they have": their career, their reputation, and their livelihood, in order to overcome government and industry wrongdoing.[3]

Caroline Snyder
Emeritus Professor
College of Liberal Arts
Rochester Institute of Technology

APPENDIX I: TEN MYTHS ABOUT BIOSOLIDS

Citizens for Sludge-Free Land (CSFL)
www.sludgefacts.org
September 6, 2013

MYTH NO. 1: For more than 2000 years industrial waste and sewage sludge have been land-applied as soil amendments. (Source: EPA[1])

FACT: The myriad hazardous industrial chemical wastes found concentrated in modern treated sewage sludges (biosolids), including pesticides, pharmaceuticals, plasticizers, flame retardants and growth hormones, to mention a few, did not even exist until recent decades.

MYTH NO. 2: Biosolids are nutrient-rich organic fertilizers. (Source: EPA[2])

FACT: It's highly deceptive to call mixtures of many thousands of industrial chemical pollutants "nutrient-rich" simply because several of the pollutants are nitrogen and phosphorus compounds found in commercial fertilizers. Biosolids produced from sewage sludges generated in industrial urban centers are undoubtedly the most pollutant-rich materials on Earth. When applied to land, industrial pollutants in biosolids reenter aquatic systems and are magnified up the food chain.[3]

MYTH NO. 3: Over 99 percent of biosolids is composed of water, organic matter, sand, silt, and common natural elements. (Source: NEBRA[4])

FACT: It's also deceptive to call mixtures of many thousands of industrial chemical pollutants "natural," especially when EPA and the biosolids industry are targeting consumers who use the words "natural" and "organic" to mean free of synthetic chemical contaminants.

MYTH NO. 4: Biosolids are essentially pathogen free. (Source: State of California[5])

FACT: Many if not most pathogenic (disease-causing) bacteria and viruses can survive treatment processes used to produce biosolids (Class A and Class B); and many dangerous pathogens, such as *Salmonella* and *Staphylococcus*, can re-grow to high levels in biosolids, which is mostly comprised of human feces.[6] New research indicates that sewage sludge treatment facilities are actually breeding grounds for antibiotic-resistant pathogens.[7]

MYTH NO. 5: Infectious prions will not survive wastewater treatment and therefore are not present in land-applied biosolids. (Source: U. Arizona[8])

FACT: The latest research shows that prions survive wasterwater treatment processes.[9]

MYTH NO. 6: Biosolids are not sources of pathogens or toxicants. Sludge syndrome is a somatic disease triggered by biosolids odors and by fears raised in the community and through the media. (Source: Mid-Atlantic Biosolids Association[10])

FACT: Odors from biosolids are a warning that the material is emitting disease-causing pathogens and biological toxins, e.g., endotoxins. Peer-reviewed scientific studies have demonstrated that resulting health effects are not imagined but real.[11]

MYTH NO. 7: Allegations of health problems linked to biosolids exposure are urban myths. (Source: NEBRA[12])

FACT: Many hundreds of sludge-exposed rural neighbors have reported chronic respiratory, skin and gastrointestinal conditions consistent with exposures to the types of chemical and biological contaminants found in biosolids. The relationship between land application of biosolids and such adverse health effects has been documented in valid scientific studies, including the peer-reviewed scientific literature.[13]

MYTH NO. 8: Treatment breaks down most organic chemical pollutants. (Source: NEBRA[14])

FACT: EPA's 2009 Targeted National Sewage Sludge Survey of 74 sewage treatment plants in 38 states, which sampled 145 industrial chemical pollutants, found them in every sample.[15] Their concentration ranges often topped ppm-levels and higher, exceeding concentrations considered safe in drinking water by orders of magnitude. Moreover, the breakdown products from organic chemical pollutants are often more harmful than the parent compounds.[16]

MYTH NO. 9: Biosolids contaminants are tightly bound to soil and do not become bioavailable. According to Rufus Chaney, "You can put enough heavy metals in the soil to kill the crop but that crop is still safe for human consumption." (Source: USDA[17])

FACT: EPA and the USDA buried studies demonstrating heavy metals in biosolids exceeding current levels permitted by EPA caused liver and kidney damage in farm animals grazing on fields treated with biosolids.[18] After EPA promulgated the current sludge rule in 1992, it worked with the Water Environment Federation to establish the "National Biosolids Public Acceptance Campaign." EPA's Office of Inspector General investigated EPA's efforts to silence Dr. David Lewis, one of its top scientists who documented adverse health effects, and concluded that EPA could not assure the public that land application of biosolids is safe.[19]

MYTH NO. 10: US sludge regulations (the 503s) are completely protective, based on science and valid risk assessment models. (Source: NEBRA[20])

FACT: A 1999 Cornell Waste Management Institute paper concluded that the 503s do not protect human health, agriculture, or the environment.[21] The 503s regulate only nine metals plus inorganic nutrients (N, P). Even though industry can legally discharge any amount of hazardous waste into sewage treatment plants, the rules are based on chemical-by-chemical risk assessment which ignores the effects of mixtures and interactions. The 2002 NRC biosolids panel recognized this and concluded that "it is not possible to conduct a risk assessment for biosolids at this time (or perhaps ever) that will lead to risk-management strategies that will provide adequate health protection without some form of ongoing monitoring and surveillance . . . the degree of uncertainty requires some form of active health and environmental tracking.[22]

APPENDIX II: BIOSOLIDS CADMIUM DATA PRE- AND POST-1993

Dramatic Change in Data

Mid Month Estimated 95% UCL Cadmium

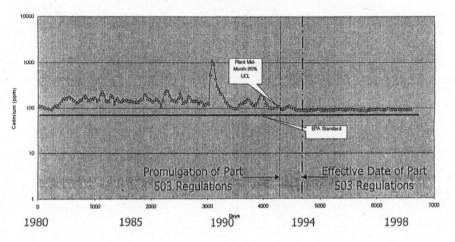

APPENDIX II: BIOSOLIDS CADMIUM DATA PRE- AND POST-1993

Dramatic Change in Data

APPENDIX III: SOME THINGS NEVER CHANGE

Keynote Address by David L. Lewis, PhD[1]
Annual March to the Athens–Clarke County Landfill
Billups Grove Baptist Church
Pastor Ben Kenneth Willis, First Lady Artherlene Willis
May 2012

When I graduated from high school and came to the University of Georgia (UGA) in 1966, Chester Davenport had just become the first African American law student to graduate from UGA. Two years later, I took a part-time job at the "Water Lab" on College Station Road, which was under the US Department of Interior. My favorite duty was going on field trips to sample rivers in North Georgia.

A lab tech named John worked in my section. Whenever we stopped to get drinks and snacks, John would sit in the car. It was many years later when I asked one of the supervisors why that was. "Because blacks weren't welcome back then," I was told. It had never entered my mind. I wasn't raised that way, and neither were my children. John passed away a number of years ago, and I still miss him.

In 1970, Congress created the US Environmental Protection Agency (EPA). Its first priority was to clean up America's polluted waterways. Our lab became part of EPA's Office of Research & Development. I earned a bachelor's degree in microbiology at UGA in 1971, and continued on to get my Ph.D. in ecology. EPA gave me a permanent position as a research microbiologist, and I had my own laboratory. Becoming a scientist and having my own lab was everything I dreamed of since I was a small child.

To clean up the water, President Jimmy Carter undertook the nation's largest public works program, which was to build sewage treatment plants in every city and town across America. Their purpose was to remove toxic pollutants from the water and concentrate them in sewage sludge, mainly for disposal in landfills. When President Carter announced the program to Congress in 1977,[2] he warned, "We need to be sure that sewage projects supported by federal money do not create additional environmental problems .... We also must ensure that the systems are operated properly...that there is an effective pretreatment program to remove harmful industrial wastes from these systems, and that we are carefully considering alternative solutions."

Shortly after President Bill Clinton began his first term in 1993, EPA passed the 503 sludge rule over the objections of scientists at EPA laboratories across the county. The rule allowed cities to just treat their sewage sludge with lime, or by other processes that do nothing to remove heavy metals and persistent organic chemicals. The toxic sludge could then be spread on farms and forests and school playgrounds as "fertilizer."

EPA's sludge rule regulates only nine of twenty-seven heavy metals found in sewage sludge. None of the toxic organic chemicals it contains are regulated, or even monitored. Not even priority pollutants, including pesticides, pharmaceuticals, and plasticizers, are regulated in sewage sludge. It makes no sense. What everyone agreed was harmful to public health and the environment in the water has suddenly become environmentally beneficial when spread on the land.

Under the direction of Dr. Rosemarie Russo, the Athens EPA laboratory led the opposition within EPA's Office of Research & Development. But EPA passed the 503 sludge rule anyway, promising to support more research to make it safe. To do the research, EPA established a cooperative agreement with the Water Environment Federation, a leading wastewater industry trade association.

The Water Environment Federation, or WEF as it was known, decided sewage sludge would be called biosolids, and carried out the National Biosolids Public Acceptance Campaign funded through EPA using congressional earmarks. Together, EPA, the US Department of Agriculture, and the WEF hand-picked scientists they wanted to support at land grant universities, including the University of Georgia. These scientists and their universities were heavily funded to publish research supporting EPA's sludge rule as being protective of public health and the environment.

By the late 1990s, reports of adverse health effects started showing up in local newspapers across the United States and Canada. Skin lesions often developed in people who contacted the material. Residents near land application sites reported burning eyes, burning lungs, and difficulty breathing when exposed to dusts blowing from treated fields. People who couldn't afford to move away developed chronic infections and permanent scarring of the lungs. Some died. You don't need me to tell you that a number of people here at Billups Grove Baptist Church are suffering from these same illnesses, which started when Athens–Clarke County began stockpiling bio-solids behind the church and spreading it on the fields near where you live.

In 1996, as part of my official EPA duties, I began investigating public health problems linked to biosolids. Soon, Dr. Russo started getting calls from EPA headquarters. One of her bosses said, "Put a muzzle on Lewis." Another asked, "Does he have some kind of death wish?" Dr. Russo refused to stop my research, and was soon ordered to step down. Fortunately, Congress held hearings, and EPA backed down with regard to removing Dr. Russo. Eventually, EPA took away all of my funding and transferred me to UGA to await termination. There, I continued to document problems with biosolids using my own personal funds.

To discredit my research, EPA funded the University of Georgia in 1999 to conduct a research project related to two dairy farms I was investigating near Augusta, Georgia. Hundreds of cows had died after eating hay fertilized with biosolids. Several workers spreading the hay got sick; one was rushed to the hospital. This EPA-funded project was published in 2003. The authors at UGA and EPA concluded that heavy metals and nitrogen in grasses grown on Augusta's sludge "did not pose a risk to animal health." The university issued a national press release saying, "Some individuals have questioned whether the 503 regulations are protective of the public and the environment. This study puts some of those fears to rest."

The same year that the UGA study was published, EPA terminated me for documenting illnesses and deaths linked to biosolids. Dr. Russo provided the following public statement: "Dr. Lewis' involuntary termination over his research articles was not supported by the local lab management in Athens. He was an excellent researcher and an asset to EPA science." EPA claims that science proves that biosolids are safe. But this "science" was created by paying scientists to support EPA's sludge rule, and firing scientists who don't. It's not real science.

The dairy farmers and I filed several lawsuits over EPA's and UGA's efforts to cover up problems with Augusta's biosolids, including with the study published in 2003. In 2008, Judge Anthony Alaimo of the Southern District of Georgia ruled that key data EPA provided in the study were "invented." One of the plant managers in Augusta, who was the original source of these data, admitted under oath that he sat down at his computer one day in 1999 and fudged two decades worth of data. He made the levels of heavy metals in Augusta's biosolids appear to be much lower after the 503 sludge rule passed in 1993.

Documents that turned up in a "qui tam" lawsuit, which the dairy farmers and I filed on behalf of the US government, uncovered something that was even more startling. Authors of the EPA-funded study failed to report that UGA's School of Veterinary Medicine had performed autopsies on two cows from one of the fields treated with Augusta's biosolids. The pathologist found that the cows' kidneys were probably damaged by one of the heavy metals found at elevated levels in the biosolids, and concluded that the kidney damage could have compromised their immune systems and caused them to die from a rare type of infection.

I'm sure it comes as no surprise to you that inner-city neighborhoods and economically and educationally disadvantaged communities across America are prime targets for disposing of sewage sludge. So let me close by paraphrasing a short section in one of my articles.

In 2005, the USDA and the Kennedy Krieger Institute at Johns Hopkins University published the results of experiments in which lead-contaminated biosolids were added to soils in predominantly African American neighborhoods in Baltimore, Maryland. According to my analysis, which was requested by Maryland's Office of Civil Rights, the levels of lead in the mixtures of soil and biosolids to which children were exposed in this experiment exceeded CDC safety limits. Moreover, the study did not follow normal experimental protocols for testing residents and their homes for lead levels.

The Kennedy Krieger Institute and USDA used this and other studies to conclude that biosolids can prevent children from getting lead poisoning. In a similar study involving lead paint, the KKI was sued by parents whose children apparently developed lead poisoning. A Maryland appeals court likened the research to the infamous Tuskegee syphilis study, and to Nazi medical experiments in World War II.

In 2005, EPA and state health officials held a public "health fair" in an African American community in Louisiana to instruct residents in personal hygiene after an outbreak of staph infections. Residents developed boils when biosolids were applied to sugar cane fields where they lived, but the state health department dismissed biosolids as having any role in the outbreak.

In a report summarizing their findings, EPA and state health officials never mentioned peer-reviewed research articles my coworkers and I had published concerning biosolids and outbreaks of staph infections. In our research, we found that one-fourth of the cases we studied at ten land application sites, including several deaths, involved staph infections associated with chemical irritants in biosolids.

I wonder what my old friend John would have to say about all of this if he were still alive today? What would he think about Congress creating EPA to clean up the water, only to spread all the toxic pollutants in inner-city neighborhoods and rural communities where people are too poor to do anything about it? What would he have to say about President Carter's long forgotten warning to Congress not to let wastewater treatment plants just end up solving one problem to create another?

The world has come a long way since John and I traveled the back roads of Georgia collecting water samples in the 1960s. I'm sure he would have been proud to see President Obama swept into the White House by a nation hungry for more change. But if he were here with us today, I think he would just look around, smell the air, and say, "Some things never change."

Judge Clay Land of the Middle District of Georgia dismissed our qui tam lawsuit on a technicality and ordered the dairy farmers and me to pay over $61,000 in court fees demanded by EPA and the University of Georgia, which they had promised to drop if we would agree not to talk about the case publicly. But we're not going to shut up. The farmers lost their dairy farms, and I lost my job. So we don't have the money to pay either.

In 1776, just nine years before the University of Georgia was chartered to become the newly formed nation's first university, the signers of the Declaration of Independence wrote, "We hold these truths to be self-evident, that all men are created equal, that they are endowed by their Creator with certain unalienable Rights, that among these are Life, Liberty and the pursuit of Happiness."

The government took away these rights when it decided to dump the nation's toxic wastes on the most poor and defenseless segments of its population, and spread them around their homes and churches. And if the government threatens citizens who speak out with unbearable court costs to gain their silence, it assaults one of our most sacred rights under the Constitution, which is the right to speak freely.

We live in the shadows of a great university that was once known for discriminating against people of color. *Some things never change.* Research at the

University of Georgia is funded, in part, by a government that once carried out the infamous syphilis experiments on African Americans in Tuskegee, Alabama. *Some things never change.* Today, we are gathered at Billups Grove Baptist Church to tell the federal government, the University of Georgia, and Athens–Clarke County, *It's time for change.*

APPENDIX IV: SLUDGE MAGIC VERSUS NATIVE AMERICAN BELIEFS

Chapter 5, "Black Magic," deals with the deceptive nature of EPA's lead abatement studies in African American communities, which it uses to argue that land application of sewage sludge (aka biosolids) reduces risks of lead poisoning. With respect to Native Americans, EPA's promotion of pollutant-rich biosolids as being beneficial to land is more than just deceptive. It's disrespectful of the reverence Native Americans have for their ancestors and the lands they once inhabited. And the agency's history of covering up adverse health effects only perpetuates the kind of treatment American Indians have received from the US government since it removed them from their fertile lands and gave them barren wastelands in return.

A case in point is the towering pile of sewage sludge called "Mount San Diego," which sits on the Torres Martinez Indian Reservation in Southern California. Members of numerous Native American tribes, environmental activists, and farm workers blockaded entrances to the site for fourteen days in October 1994 in an effort to shut down the operation. According to a fact sheet published by the Water Environment Federation (WEF), the reservation's residents complained of "thick dust and clouds of flies," along with adverse health effects, including allergic reactions, headaches, and nausea.[1] To put down their complaints, the WEF pointed to tests conducted by federal and state agencies that purportedly showed that "composting at the Torres Martinez reservation caused no environmental damage or threat to human health." It was yet one more example of the injustices EPA created from 1992 to 1999 when its Office of Water funded the WEF to investigate "biosolids horror stories" and issue "fact sheets" to inform the public of "what really happened."[2] The WEF, of course, never found any public health problems with land application of biosolids.

The enduring connection Native Americans have with the lands their ancestors freely roamed is eloquently expressed in popular versions of a speech attributed to Chief Seattle, a Suquamish Indian who sold his tribe's land to the US government in the mid-1800s. The following excerpt serves as an example:[3]

> *Every part of this soil is sacred in the estimation of my people. Every hillside, every valley, every plain and grove, has been hallowed by some sad or happy event in days long vanished. Even the rocks, which seem to be dumb and dead as they swelter in the sun along the silent shore, thrill with memories of stirring events connected with the lives of my people, and the very dust upon which you now stand responds more lovingly to their footsteps than yours, because it is rich with the blood of our ancestors, and our bare feet are conscious of the sympathetic touch.... At night when the streets of your cities and villages are silent and you think them deserted, they will throng with the returning hosts that once filled them and still love this beautiful land.*

A quarter of a century ago, the pastor of our small community church and I purchased a tract of wooded land in Oconee County, Georgia, that includes a stretch of Little Rose Creek. When we first looked at the property, he asked whether I thought it may have contained any Indian artifacts. I pointed to a small depression visible in a distance as a likely spot. As we approached it, we noticed a small, red projectile point laying in a small pocket of eroded clay. Since then, I'm continually reminded of the original inhabitants who took care of that land long before we built the church and my family picked out a spot to live near the creek. Having volunteered to help at archeological sites, I've learned to appreciate the artistic skill and craftsmanship exhibited by objects Native Americans fashioned from clay and stone. Each of them has a story to tell, and none tug at the heart more than some I've found along the creek bed.

Coming across evidence of daily life left behind by these original inhabitants is different from looking at artifacts in museums. There, I have no personal connection with the people who created the objects. But when I see them on the land where I live, I can almost sense the presence of those who lived there before me. Their tools and works of art have something important to say about the land, and the people who lived there in the distant past.

One shard of Indian pottery lying among the rocks in the creek was particularly heartrending. Under the rippling water I could see that its clay surface also formed ripples, which radiated from a tiny hole in the design. It was once part of an elaborately decorated pot approximately fifteen inches in diameter. As I lifted the fragment out of the water, I noticed that the hole represented the eye of a fish breaking the water's surface, as if caught on a line. Looking back, I can't help but think about the "happy event in days long vanished" that it represents in the life of someone from another era—someone whose ancestors and descendants loved the creek's clear, flowing water and surrounding woodland for many generations, whose people will never get to experience such happiness in that place again. To me, it tells the story of a way of life that has been lost, a door forever shut.

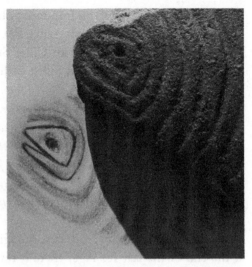

Pottery shard with fish-head design. Courtesy David Lewis.

Rock bearing a Christian cross found among Indian artifacts. Courtesy David Lewis.

Nearby, I noticed a flat rock about one-and-a-half inches thick with beveled ends that appeared to have a petroglyph that told a different story. The beveled underside fit comfortably in the palm of my hand. Its face was flat like a tablet, and bore what looked like the chiseled image of a Christian cross about 2.5 × 5 inches long. Perhaps some early homesteader used it to mark a small grave. Or, because it was nestled among Indian artifacts, maybe one of the last remaining Native Americans living in the area gouged it with an iron tool as he sat on the banks and pondered the strange God of his enemies. Later, as I traced the grooved lines with my finger, I thought about a part of Chief Seattle's speech that contrasts the earthly handiwork of the Great Spirit with that of Abraham's God.

> *The white man's God cannot love our people, or he would protect them. They seem to be orphans who can look nowhere for help.... Your religion was written upon tablets of stone by the iron finger of your God so that you could not forget.... [We] could never comprehend or remember it. Our religion is the traditions of our ancestors... written in the hearts of our people.*

In the same area, the creek cuts though a layer of blue clay that formed at the end of the last Ice Age. There it revealed the broken mid-section of a Paleolithic blade or projectile point. Thin, and perfectly symmetrical, it proved that Indians had occupied this spot since the very earliest times of human history in North America. The story it tells is one of impending doom for all humanity. About twelve thousand years ago, these amazing people barely survived an asteroid or comet that impacted the Laurentide Ice Sheet covering what is now Canada. Other than bison, no other large mammals in North America escaped the fire and burning acid that rained down, and the prolonged darkness and deep freeze that followed. Yet, these earliest inhabitants were able to pick up their belongings and make their way from the Eastern Seaboard to find shelter in the mountains out West. It was no less of an achievement than our sending humans to the moon, and bringing them back alive.

Instead of acknowledging their great heritage, the managers of EPA's biosolids program treat Native Americans not unlike their ancestors were treated under the Indian Removal Act of 1830. Their ancient lands are being destroyed with municipal and industrial wastes. And their descendants who complain of poisoned soils and foul air on their reservations are just seen as obstacles standing in the way of the government's plans to pollute those lands as well. Within EPA's biosolids program, there is no reverence at all for the land, or any respect for the people who live on it. Until that changes, there remains no hope, neither for the land nor the people. Yet EPA's biosolids program is just a microcosm of our federal government as a whole, and the governments of other highly industrialized nations. It would be far easier for them to put people on the moon than enable their populations to survive an event such as the one that befell Native Americans at the end of the last Ice Age. It shows how far removed modern technology is from integrating civilization with the natural world. And yet we all live on the brink of such events, which inevitably and suddenly come forth from the earth below and the heavens above from time to time.

One of the most rare and precious resources that our government held in its hands a century or two ago was the knowledge and understanding of our natural environment possessed by Native Americans. But, instead of protecting Native Americans, it set out to obliterate them. In doing so, it committed us all to a less secure future. Even today, we should take stock of what is left of this resource, and enlist Native Americans in an effort to figure out how to do what their ancestors did. Their enduring reverence for the land is just what EPA policymakers need to protect it for future generations to enjoy, and to serve as a safe haven when future cataclysmic events threaten our survival as a species once again.

APPENDIX V: SENATE TESTIMONY PREPARED BY DAVID LEWIS

Briefing on "Oversight on the State of Science and Potential Issues Associated with EPA's Sewage Sludge Program"
September 11, 2008

TESTIMONY OF DAVID L. LEWIS, PhD[1]
Director, International Center for Research on Public Health and the Environment, www.researchcenter.uga.edu

Visiting Scientist, Department of Marine Sciences,
University of Georgia Athens, Georgia 30602, DavidL@uga.edu

Chairman Boxer, Ranking Member Inhofe, and honorable members of the committee, thank you for the privilege of testifying today about my research on sewage sludge at the U.S. Environmental Protection Agency.

OPENING STATEMENT OUTLINE
"EPA's handling of the 503 sludge rule"

- EPA-ORD identified six scientific weaknesses, which have existed since EPA's sludge program began
- EPA-OW reneged on its promise to fix these weaknesses
- OW/ORD silenced scientists who tried to fix these weaknesses
- Witness A. McElmurray and I would not be here today had OW worked with ORD to fix these weaknesses
- EPA needs to fix OW/ORD
- My written testimony answers twelve questions concerning how the US Environmental Protection Agency (EPA) has handled the sewage sludge issue.

1. What is sewage sludge?

Sewage sludge is the semi-solid material that settles out at municipal waste-water treatment plants when water containing effluent from municipal and industrial sources is purified. It consists mostly of human feces and contains a variety of chemical and biological contaminants, which enter municipal sewerage systems from various sources, including homes, businesses, hospitals, and factories. Some municipalities also direct stormwater runoff into sewer lines for treatment at wastewater treatment plants.

2. What are biosolids?

In 1978, John Walker at EPA's Office of Water (OW) proposed to associate deputy assistant administrator Henry Longest that EPA support the use of sewage sludge as an inexpensive fertilizer for agriculture. Walker wrote, "The application of some low levels of toxic substances to land for food crop production should not be prohibited; rather, it should be controlled by proper rates of sludge/toxic application, soil management, etc."[2] Longest and his superiors supported the idea and EPA began developing guidelines for processing sewage sludge into fertilizer, which came to be known as "biosolids."[3]

Based on levels of certain bacteria or viruses called "indicator" pathogens, which are used to *indicate* the presence of pathogens, processed sewage sludge is distributed as either Class A or Class B biosolids. (Testimony Appendix I)

3. Why did I conduct research on sewage sludge at EPA?

In 1996, I wrote a commentary for *Nature* titled "EPA Science: Casualty of Election Politics.[4] In it, I discussed a number of areas where EPA regulations involve microbiological processes but fail to incorporate this area of science into the regulations. Relatively few microbiologists worked for EPA at the time.

When a local reporter with the *Atlanta Journal-Constitution* asked me for a worst-case example, I put the question to other scientists working at EPA research laboratories across the country. Their suggestion was the 503 sludge rule, in which EPA's Office of Water had not done a risk assessment for pathogens (disease-causing microorganisms). The newspaper ran a full-page article about the commentary and focused on the sludge rule. I filed several successful complaints with the Department of Labor to fend off retaliations.

To settle one of the complaints in 1998, my top career EPA manager in the Office of Research & Development (ORD), Henry Longest, proposed that I be assigned to the University of Georgia (UGA) under the

Intergovernmental Personnel Act (IPA) to continue earlier work that I had done concerning infection control in dentistry. As a member of the graduate faculty at UGA, I had collaborated with researchers at dental and medical schools in California and Missouri to investigate an outbreak of human immunodeficiency virus (HIV), the AIDS virus, in a Florida dental practice.

Working after hours with a dentist treating AIDS patients in Atlanta, we discovered that the AIDS virus could escape disinfection by hiding in lubricants in dental devices used for drilling and polishing teeth. Live HIV, we found, leaked back out in amounts that were high enough to readily infect human blood cells (lymphocytes) when the devices were reused. We published the work in *Lancet* and *Nature Medicine*, and our findings convinced public health organizations worldwide to issue new infection control guidelines for dentistry.[5]

For the IPA assignment Longest offered me, I proposed organizing an epidemiological study of endoscopy-related hepatitis C (HCV) cross-infections in Egypt where that virus is highly prevalent.[6] Flexible endoscopes used in colorectal cancer screening and many other important diagnostic and therapeutic procedures are not designed with infection control in mind. Approximately 80 percent of these devices used in the United States and elsewhere contain internal air and water channels that are too narrow to brush out visible amounts of blood, tissue, and feces that collect inside them.

IPA rules required that I apply my research on flexible endoscopes to EPA's mission. There was widespread support among scientists within ORD to address the issue of pathogens in sewage sludge; therefore, local EPA managers agreed that I could apply my research on infection control in endoscopy to EPA's mission by investigating risks posed by pathogens in sewage sludge. ORD's concern was that exposing the US population to pathogens in municipal wastes could potentially provide an open door to emerging infectious diseases. So, I intended to fill in some of the gaps in scientific data that were missing as a result of EPA not having done a pathogens risk assessment for the 503 sludge rule.

To accomplish my research on sewage sludge, I assembled a team of scientists at the University of Georgia, including a professor of medical microbiology, to study anecdotal reports of illnesses and three deaths reported by residents living near land-application sites. We collaborated with a physician in California treating children exposed to sewage sludge, and I paid for all of the research out of my personal funds. Altogether, we studied ten land application sites in the United States and Canada.

We were interested in investigating these anecdotal cases even though they were insignificant in number compared with, for example, the numbers of deaths attributed to auto accidents, smoking, and other major public health issues. Whether these cases would turn out to indicate a significant public health problem depended on what the actual number of cases was and whether there were any indirect effects involving larger groups.

Dentistry is one example in which even a few documented cases may have been the proverbial tip of the iceberg. For reasons that are not yet fully understood, the hepatitis C epidemic in the United States began to precipitously collapse in 1988. The downturn in HCV cases, which occurred mostly among injection drug users, coincided with the first suspected case of HIV infection in dentistry. In response, the American Dental Association had undertaken a national campaign to improve infection control. Hepatitis B also dropped dramatically, while hepatitis A and other non-bloodborne infections did not.

Any reduction in sporadic cases of dental infections and the collapse of the HCV epidemic may be entirely coincidental. However, HCV is primarily transmitted by sharing blood-contaminated devices, and drug addicts did not suddenly stop sharing needles in 1988. Our research suggests that when inefficient mechanisms of transmission, e.g., those involving poorly disinfected dental devices, become commonplace, they may actually play an important role in driving major epidemics. The reason is that they could be moving infectious agents around within the general population and introducing them to high risk groups engaged in more efficient mechanisms, e.g., needle sharing.

4.   How are biosolids applied to land, and what aspect of land application did I study at EPA?

Processed sewage sludge, or biosolids, is used as a fertilizer or soil conditioner in agriculture, forestry, mining reclamation, home gardens, and a variety of other uses. It is spread on the surface of land and incorporated in soil in liquid, semi-solid, and solid (e.g., pellet) forms. Liquid biosolids are also applied by sub-surface injection.

Some of my colleagues at ORD were concerned with aerosols generated when liquid sewage sludge is sprayed on land. By contrast, I focused on dry organic dusts picked up by winds. Such dusts can allow pathogens trapped inside to remain viable even while traveling great distances. (Testimony Appendix II)

5.   What contaminants do biosolids contain, and which ones did I study at EPA?
EPA estimates that there maybe trace amounts of as many as sixty thousand organic chemical contaminants in sewage sludge, including priority pollutants regulated under the federal Clean Water Act. Organic contaminants include, for example, pesticides, petroleum byproducts, antibiotics, and other pharmaceuticals. Biosolids also contain a wide variety of inorganic pollutants. These pollutants include heavy metals, such as mercury, cadmium, molybdenum, arsenic, lead, thallium, and chromium. Biosolids can also contain other inorganic pollutants, such as nitric, hydrofluoric, and sulfuric acids.

Generally speaking, any chemical contaminants present in wastewater originating from municipalities and local industries are also likely to be present in biosolids, and often at much higher concentrations. Biosolids also contain parasites and disease-causing bacteria, fungi, protozoa, and viruses. My research at EPA focused on chemical contaminants in sewage sludge dusts that could irritate the skin, eyes, mucous membranes, and respiratory tract and possibly lead to infections, especially staphylococcal infections.

6.   How are biosolids regulated, and which regulations did my research at EPA involve?
Sewage sludge, or biosolids, is regulated under 40 CFR, Part 503 (the "sludge rule" or "503 rule"). The 503 rule sets limits on levels of indicator pathogens and certain inorganic chemicals, including nine heavy metals (arsenic, cadmium, copper, lead, mercury, molybdenum, nickel, selenium, and zinc) and certain nutrients, e.g., nitrates. If processed sewage sludge fails to meet Part 503 criteria, Toxicity Characteristic Leaching Procedure (TCLP) tests must be performed to determine whether the waste is hazardous. If so, it must be disposed of accordingly under the Resource Conservation and Recovery Act (RCRA). My research related most directly to the 503 sludge rule.

7.   What are the scientific weaknesses in EPA's sludge rule, and which ones did I address in my EPA research?
When the 503 rule was promulgated in 1993, ORD identified six areas in which research was lacking: (1) monitoring of surface and groundwater to understand how sewage sludge pollutants are transported and chemically transformed in the environment; (2) characterizing the variability of land application practices in actual practice; (3) determining the bioavailability of sewage sludge pollutants for uptake by plants and animals; (4) assessing

the ecological effects of chemicals and pathogens in sewage sludge on wild-life and other plant and animal communities; (5) determining the distribution and variability in the concentrations of sewage sludge pollutants and their capacities to remain bound to the sludge; and (6) understanding long-term changes at land application sites, including, for example, changes in soil pH, land use, and the capacity for sewage sludge to bind chemical contaminants.

Upon investigating a complaint that I filed regarding OW's failure to address the six scientific weaknesses identified by ORD, EPA's Office of Inspector General (OIG) audited the program and issued a report in 2002.[7] The Inspector General found that OW failed to follow through with the research support it promised, and, as a result, EPA could not assure the public that land application is safe. The National Academy of Sciences also determined other scientific weaknesses in the 503 rule in 1996 and 2002, such as EPA's lack of a risk assessment for pathogens, a lack of epidemiological studies, and a lack of any understanding concerning how contaminants in sewage sludge may interact.

My work at EPA specifically addressed weaknesses number 1, 3, 4, 5, and 6 above, which were identified by ORD in the preamble to the 503 rule. My work also involved issues addressed by the NAS, such as investigating anecdotal cases of residents and workers reporting adverse health effects, and studying the interactions of chemicals and pathogens in sewage sludge.

8.   What were the main areas of scientific controversy, and how did my research relate to them?

- Organic pollutants: EPA does not regulate organic chemicals, including priority pollutants, in biosolids. Land application proponents argue that all potentially harmful organic chemicals in biosolids, such as pesticides, petroleum byproducts, and pharmaceuticals, are present only at very low levels and are unavailable for uptake by plants and humans. They also argue that a number of chemicals that were of concern, e.g., chlordane, have now been banned or restricted and should no longer present a problem.
- Heavy metals: EPA excludes some potentially toxic heavy metals occasionally found in sewage sludge, including, for example, thallium, antimony, and chromium.

• Pathogens: Disease-causing organisms found in sewage sludge include parasites and many kinds of bacteria, viruses, fungi, and protozoa. These organisms exhibit different levels of resistance to treatment processes designed to reduce their numbers, and the effectiveness of different treatment processes is uncertain.

In my EPA research, my coworkers and I studied residents living near land application sites. We included both affected and unaffected residents, that is, those who reported symptoms and those who did not. Based on dose-response curves and other data, we found an association between gastrointestinal, respiratory, and skin-related complaints and dusts blowing from land application sites. (Testimony Appendix III)

9. Did my research at EPA, or any other research, ever determine whether biosolids are safe?

No. Without knowing which potentially harmful contaminants are present, and at what concentrations, and how they interact as mixtures, and without performing any toxicity tests, epidemiological studies, or carefully designed cause-and-effect studies, it is impossible to know whether any particular batch of biosolids presents any significant risk to public health or the environment.

10. Did my research at EPA, or any other research, ever prove that sewage sludge applied under the 503 rule has harmed public health or the environment?

No. There are only two cases proving that land-applied sewage sludge has harmed public health or the environment. These are the McElmurray and Boyce cases, in which cattle on two dairy farms in Georgia were poisoned by high concentrations of cadmium, molybdenum, arsenic, thallium, chlordane, and other hazardous wastes in sewage sludge produced by the City of Augusta. None of the sewage sludge applied in those cases ever complied with EPA's applicable guidelines, including 40 CFR, Part 503, and its predecessor, 40 CFR, Part 257.

The peer-reviewed scientific literature contains three studies concerning the potential effects of sewage sludge on residents living near land application sites. (Testimony Appendix III) One is the 1985 Dorn study in Ohio, which found no adverse health effects. Another is our 2002 study of residents living near ten land-application sites in the US and Canada. We found a dose-response association between gastrointestinal, respiratory, and skin-related diseases and the distances residents lived from fields treated with sewage sludge. The third study, which was conducted in Ohio and published in 2007, supported our results. However, none of these studies are conclusive.

The 2002 National Academy of Sciences report, "Biosolids Applied to Land: Advancing Standards and Practices," concluded that there is no documented evidence that biosolids applied under EPA's 503 sludge rule has ever harmed public health or the environment. Our peer-reviewed studies were deleted from the final version of this report without the approval of the NAS panel.[8] Also, the Augusta cattle cases were dismissed in the report based on data that EPA's Office of Water provided to the panel, which US District Judge Anthony Alaimo recently ruled were "fudged," "fabricated," and "invented."[9]

While I believe that our EPA-approved, peer-reviewed scientific research provided credible evidence, which should not have been deleted from the NAS report, I do not believe that our research proved that sewage sludge applied under EPA's 503 rule has caused gastrointestinal, respiratory, and skin-related diseases. *Nature* accurately described our research findings in a recent editorial as having established "an association between these factors" and that "recent research underscores those findings."[10]

11. How did EPA respond to my research findings concerning land application of sewage sludge?
Local EPA managers in Athens, Georgia, fully supported my research on sewage sludge and approved all of my research plans. Nevertheless, Henry Longest and other senior EPA officials involved with the agency's policies on sewage sludge alleged that my writings in *Nature* and elsewhere violated government ethics rules. They also required special procedures for my promotion and other favorable personnel actions to be approved.

I filed a number of complaints with EPA's Office of Inspector General and the US Department of Labor. Both organizations investigated and found that OW officials and others had retaliated and/or acted improperly concerning the content of my peer-reviewed research articles and other scholarly writings. The Committee on Science in the House of Representatives held two hearings into these retaliations in 2000.[11] As a result, Congress passed the No Fear Act signed by President Bush in 2002.[12]

Then, after failing to get local EPA managers in Athens to "muzzle" me,[13] Longest offered me an opportunity to take the IPA assignment to study infection control issues at the University of Georgia. The only catch was that I had to agree to resign my EPA position as soon as I was eligible to retire. Longest had dead-ended my career at EPA, and, when UGA offered to pursue a tenured faculty position for me, I accepted his offer. Just three weeks after I signed the agreement proposed by Henry Longest, John Walker, Longest's former employee at OW, contacted land application specialist Julia Gaskin and others at UGA and offered Gaskin a research grant to conduct a study dealing with the cattle cases in Augusta.

While Walker pushed his project forward at UGA, Longest began applying the brakes to my UGA research. For example, Longest required that all actions involving my research at UGA be approved by EPA headquarters. Headquarters officials wasted no time prohibiting me from using a sophisticated mathematical model developed at UGA, which I needed to predict the complex interactions of pollutants contained in sewage sludge. They also prohibited me from collaborating with other EPA scientists who wanted to help with the research.

While my research papers were undergoing internal peer review at EPA, Walker met on two occasions with an industry representative and requested information that he could use in EPA's internal peer-review process to stop my sludge research from being published. The industry representative later emailed Walker allegations that my sludge research had never passed peer review and that my research at the University of Georgia was not approved by EPA.

Of course, only someone working at EPA would know whether EPA had approved my research plans and internally peer-reviewed my research. The industry representative was only a conduit for false information that Walker or other EPA officials provided. The head of EPA's peer-review panel, who later testified in my Labor case, called Walker's actions "disgusting" and stated that my research articles were fully approved.

Walker emailed the false allegations to an attorney defending Augusta against the McElmurray and Boyce cases. Then this attorney had Walker's letter and the allegations presented at public hearings in Georgia attended by UGA faculty members. Gaskin then took the allegations up with my UGA department head and argued that my research should be shut down. Eventually, Walker's distribution of these false allegations killed UGA's interest in hiring me and made it impossible for me to continue our research on sewage sludge.

When my four years at UGA were up in 2002, I returned to EPA just six months short of being eligible for retirement. Local managers approved my requests to investigate health complaints reported by workers spreading hay fertilized with sewage sludge from Augusta, Georgia.

Senator James Inhofe, who chaired the Environment and Public Works Committee, and Senate finance chairman Charles Grassley sent a letter to EPA administrator Christine Todd Whitman asking her to intervene in my termination, partly on the basis of national security. After the 9/11 attacks, EPA headquarters included me on a list of national experts to be contacted in the event of a bioterrorism attack on the United States. Whitman never replied, and EPA headquarters terminated me in May of 2003.

In a peer-reviewed journal article published in the *American Journal of Law and Medicine*, Professor Robert Kuehn of the University of Alabama described the tactics EPA used to end my career as an environmental scientist.[14] Earlier this year, US District Court Judge Anthony Alaimo devoted four pages of a Court Ruling to describing EPA's efforts to discredit my work at UGA, in which EPA and UGA officials used "fudged," "fabricated," and "invented" data.[15] When describing EPA's actions against me, Judge Alaimo wrote, "Senior EPA officials took extraordinary steps to quash scientific dissent, and any questioning of the EPA's biosolids program."

After retiring in 2005, my local EPA laboratory director, Dr. Rosemarie Russo, provided the following statement for public release: "Dr. Lewis' involuntary termination over his research articles was not supported by the local lab management in Athens. He was an excellent researcher and an asset to EPA science."

12.   Were it not for the opposition to my sewage sludge research, would I still be working in this area?

No. Global climate change remains my primary research interest. My 1999 research article in *Nature* concerning effects that climate change could have on risks posed by environmental pollutants only mentions sewage sludge as one example where risks associated with certain pollutants could either increase or decrease.[16]

CONCLUSIONS

In 1996, EPA officials dismissed my concerns published in *Nature* over the growing political control of EPA's science. A recent survey, however, shows that the situation has grown much worse.[17] A total of 492 EPA scientists (31 percent of survey respondents) said that they could not express concerns about EPA's work to colleagues without fear of retaliation. Twice that many (60 percent) reported a personal experience of political interference. "The proportion of EPA scientists reporting interference," according to the survey, "was highest in the agency's offices with regulatory duties and at its head-quarters, while it was lowest in the Office of Research and Development, the EPA's main research arm."

My recommendation to this committee is to focus on the problem that lies at the heart of why we are here today. The failure of senior EPA officials to properly manage the agency's sewage sludge program is only one small symptom of an organization that lacks the leadership it needs to effectively integrate common sense and good science into its regulations and enforcement activities. Given the chance, EPA's Office of Research & Development has the expertise to resolve the issues surrounding the 503 sludge rule and, I am confident, every other rule and regulation that the agency has promulgated.

Before 1993, EPA rules and regulations required the concurrence of each assistant administrator, including the assistant administrator for ORD, which was established by Congress to independently review the science behind EPA's regulations. That balance between program offices, which develop regulations, and ORD, which reviews the science behind those regulations, needs to be restored.

My hope is that what the committee learns from this briefing will be used a hammer to drive the next president of the United States to give EPA the kind of leadership it needs to adequately insulate the agency's science from outside forces, no matter how powerful or well-meaning the individuals behind those forces may be. Perhaps our way of life could accommodate the kind of EPA we have had in the past. But we can no longer afford to preserve EPA in its present state, not when the very scientists we depend on to find solutions to problems that threaten our very existence are intimidated into silence or driven out whenever they disagree with the powers that be.

Appendix I. Class A and Class B Biosolids

Untreated sewage sludge contains large numbers of disease-causing microorganisms, called pathogens, which include many different kinds of bacteria, viruses, fungi, and protozoa. To produce biosolids, various treatment processes, including lime stabilization, anaerobic and aerobic digestion, composting, and heat pelletization, are used to reduce pathogen levels.

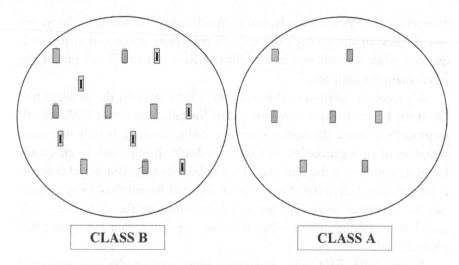

The U.S. Environmental Protection Agency established two classes of processed sewage sludge (biosolids), Class A and Class B. Class A biosolids have undetectable levels of "indicator" pathogens, such as *Salmonella* or *E. coli*, which are used to "indicate" the levels of pathogens. The most common type of biosolids, Class B, contains low levels of indicator bacteria.

The diagram above illustrates how common waste treatment processes, such as lime stabilization and composting, reduce the numbers of *Salmonella* and *E. coli* (boxes with "I"). However, some resistant pathogens, such as *Listeria, Cryptosporidium,* and *Giardia,* can still survive.[18] During storage, organic nutrients in Class A biosolids can also cause *Salmonella* and certain other pathogens to re-grow. Class A and Class B refer only to levels of indicator pathogens. They can have the same levels of chemical pollutants.

APPENDIX II. Pathogens in Dusts

Bacteria and fungi, some with the potential to cause disease in plants or animals, may be finding their way from Africa to the Americas by hitchhiking on microscopic dust particles kicked up by storms in the Sahara.[19] "This study presents evidence of early summer survival and transport of microorganisms from North Africa to a mid-Atlantic research site," says Dale Griffin of the US Geological Survey in St. Petersburg, Florida, one of the researchers on the study.

Griffin and his colleagues tested air samples on a research ship in the middle of the Atlantic Ocean during May and June 2003 to determine if airborne, viable populations of bacteria and fungi could be detected and also to see if total population counts increased with the presence of airborne desert dust.

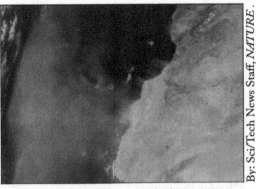

By: Sci/Tech News Staff, *NATURE*.

Microbes Hitchhike across the Atlantic on Desert Dust—From Sahara to North America

"The phenomenon known as desert-dust storms moves an estimated 2.2 billion metric tons of soil and dried sediment through the earth's atmosphere each year. The largest of these events is capable of dispersing large quantities of dust across oceans and continents. Because a gram of desert soil may contain as many as 1 billion bacterial cells, the presence of airborne dust should correspond with increased concentrations of airborne microorganisms," says Griffin.

Viable bacterial and fungal populations were collected on twenty-four of forty sampling days. The three days where the highest populations were collected corresponded with the two highest periods of dust activity as determined by the US Navy's Naval Aerosol Analysis and Prediction System Global Aerosol Model.

DNA analysis matched two of the isolates 100 percent to two dust-borne isolates previously collected from the atmosphere in Mali. One of them, a known human pathogen, has also been found in atmospheric samples in the US Virgin Islands when African desert dust was present. Additional analysis identified a number of bacteria and fungi capable of causing disease in animals and plants, including the cause of Florida Sycamore canker.[20] "It is tempting to speculate that transatlantic transport of dust could be a vector to renew reservoirs of some plant and animal pathogens in North America and could also be the cause of new diseases," says Griffin. Photo credit: NASA.

Appendix III. Selected Peer-Reviewed Scientific Articles by Lewis and Coworkers

Interactions of Pathogens and Irritant Chemicals in Land-Applied Sewage Sludges (biosolids)

By David L. Lewis, David K. Gattie, Marc E. Novak, Susan Sanchez, and Charles Pumphrey

June 28 2002

http://www.biomedcentral.com/1471-2458/2/11

Abstract

Background

Fertilization of land with processed sewage sludges, which often contain low levels of pathogens, endotoxins, and trace amounts of industrial and household chemicals, has become common practice in Western Europe, the United States, and Canada. Local governments, however, are increasingly restricting or banning the practice in response to residents reporting adverse health effects. These self-reported illnesses have not been studied, and methods for assessing exposures of residential communities to contaminants from processed sewage sludges need to be developed.

Methods

To describe and document adverse effects reported by residents, forty-eight individuals at ten sites in the United States and Canada were questioned about their environmental exposures and symptoms. Information was obtained on five additional cases where an outbreak of staphylococcal infections occurred near a land application site in Robesonia, Pennsylvania. Medical records were reviewed in cases involving hospitalization or other medical treatment. Because most complaints were associated with airborne contaminants, an air dispersion model was used as a means for potentially ruling out exposure to sludge as the cause of adverse effects.

Results

Affected residents lived within approximately one kilometer of land application sites and generally complained of irritation (e.g., skin rashes and burning of the eyes, throat, and lungs) after exposure to winds blowing from treated fields. A prevalence of *Staphylococcus aureus* infections of the skin and respiratory tract was found. Approximately one in four of fifty-four individuals were infected, including two mortalities (septicaemia, pneumonia). This result was consistent with the prevalence of *S. aureus* infections accompanying diaper rashes in which the organism, which is commonly found in the lower human colon, tends to invade irritated or inflamed tissue.

Conclusions

When assessing public health risks from applying sewage sludges in residential areas, potential interactions of chemical contaminants with low levels of pathogens should be considered. An increased risk of infection may occur when allergic and nonallergic reactions to endotoxins and other chemical components irritate skin and mucus membranes and thereby compromise normal barriers to infection.

Health Survey of Residents Living Near Farm Fields Permitted to Receive Biosolids

Archives of Environmental & Occupational Health
Vol. 62, No. 1, 2007
By Sadik Khuder, Ph.D.; Sheryl A. Milz, Ph.D.; Michael Bisesi, Ph.D.; Robert Vincent, Ph.D.; Wendy McNulty, M.S.; Kevin Czajkowski, Ph.D.
www.ohiowea.org/owea/residuals/HealthSurveyResidents08.pdf
Archives of Environmental & Occupational Health
Vol. 62, No. 1, 2007
Health Survey of Residents Living Near Farm Fields Permitted to Receive Biosolids
Sadik Khuder, Ph.D.; Sheryl A. Milz, Ph.D.; Michael Bisesi, Ph.D.; Robert Vincent, Ph.D.; Wendy McNulty, M.S.; Kevin Czajkowski, Ph.D.;

Abstract

The authors studied the health status of residents living in Wood County, Ohio, near farm fields that were permitted to receive biosolids. They mailed a health survey to 607 households and received completed surveys from 437 people exposed to biosolids (living on or within one mile of the fields where application was permitted) and from 176 people not exposed to biosolids (living more than one mile from the fields where application was permitted). The authors allowed for up to six surveys per household. Results revealed that some reported health-related symptoms were statistically significantly elevated among the exposed residents, including excessive secretion of tears, abdominal bloating, jaundice, skin ulcer, dehydration, weight loss, and general weakness. The frequency of reported occurrence of bronchitis, upper respiratory infection, and giardiasis were also statistically significantly elevated. The findings suggest an increased risk for certain respiratory, gastrointestinal, and other diseases among residents living near farm fields on which the use of biosolids was permitted. However, further studies are needed to address the limitations cited in this study.

Authors' Comment

We observed an association between respiratory, gastrointestinal, and general symptoms linked with infectious diseases and residence in homes near farm fields permitted to receive Class B biosolids. Moreover, we found a significant dose-response relationship for excessive secretion of tears, abdominal bloating, and dehydration. These findings are in agreement with the findings of Lewis et al.[21] and studies on wastewater treatment workers. However, they contradict an earlier study from three areas in Ohio, in which researchers reported no significant differences in the risk of respiratory, gastrointestinal, and general symptoms between sludge-farm residents and control-farm residents. In the Ohio study, the biosolids application rates were low, and thus exposure levels may not have been comparable to those in this study.[22]

Commentary

A High-Level Disinfection Standard for Land Applying Sewage Sludges (Biosolids)

David K. Gattie and David L. Lewis,

Environmental Health Perspectives Volume 112, No. 2, Feburary 2004

By David K. Gattie, David L. Lewis

Abstract

Complaints associated with land-applied sewage sludges primarily involve irritation of the skin, mucous membranes, and the respiratory tract accompanied by opportunistic infections. Volatile emissions and organic dusts appear to be the main source of irritation. Occasionally, chronic gastrointestinal problems are reported by affected residents who have private wells. To prevent acute health effects, we recommend that the current system of classifying sludges based on indicator pathogen levels (Class A and Class B) be replaced with a single high-level disinfection standard and that methods used to treat sludges be improved to reduce levels of irritant chemicals, especially endotoxins. A national opinion survey of individuals impacted by or concerned about the safety of land-application practices indicated that most did not consider the practice inherently unsafe but that they lacked confidence in research supported by federal and state agencies.

APPENDIX VI: SENATE TESTIMONY PREPARED BY ANDY MCELMURRAY

Briefing on "Oversight on the State of Science and Potential Issues Associated with EPA's Sewage Sludge Program"
September 11, 2008

TESTIMONY OF ROBERT A. (ANDY) MCELMURRAY, III[1]
R. A. McElmurray & Sons, Inc.
2010 Brown Road, Hephzibah, Georgia 30815
McElmurray@aol.com

Chairman Boxer, Ranking Member Inhofe, and honorable members of the committee, thank you for the privilege of testifying today about the destruction of our dairy farm business by hazardous wastes in sewage sludge, which was land-applied by the City of Augusta, Georgia.

Cattle Deaths, Milk Contamination

My name is Andy McElmurray, and with me today is my attorney, Ed Hallman of Decker, Hallman, Barber & Briggs in Atlanta, Georgia. Mr. Hallman has led a team of attorneys and experts for the last ten years in an effort to recover compensation for the destruction of my family's dairy farm business, which resulted from hazardous wastes in Augusta, Georgia's sewage sludge. My testimony addresses the history of sewage sludge applications to my family's farmlands. The City of Augusta invited us to participate in its land application program and assured us that the sewage sludge was safe for growing forage crops to feed to our dairy cattle.

We began receiving sewage sludge applications in 1979 and continued until 1990. On our farm, we grew forage crops to feed to our dairy cattle, and we grew row crops as well. In 1998, after hundreds of head of cattle sickened and died, we learned that Augusta's sewage sludge contained extremely high levels of hazardous wastes that were toxic to dairy cattle.

Another prize-winning dairy farm in the area owned by the family of Bill Boyce was hit even harder, and the owners had to abandon the dairy farm business altogether. Our families, who have farmed our land for three generations, have lost tens of millions of dollars in property value, lost property and agricultural products.

For over two decades, the City of Augusta, Georgia, failed to enforce federal and state regulations requiring local industries to treat hazardous wastes before discharging them into the city's sewers. The city also fudged, fabricated, and invented data required under the Clean Water Act to make its sewage sludge appear to qualify as "Class B biosolids." The bogus fertilizer ended up sickening and killing hundreds of dairy cows on the two dairy farms.

Milk samples collected from one of our farms still using forage grown on lands which received sewage sludge contained high levels of heavy metals and other sludge contaminants. Additional samples of milk pulled from shelves in grocery stores in Georgia and surrounding states also contained some of the same heavy metals at levels exceeding EPA's safe drinking water standards.[2] Unsafe levels of heavy metals in various samples included thallium, a rat poison toxic to humans in very small doses.

Earlier this year, US District Court Judge Anthony Alaimo rejected Augusta's fabricated data and ruled that the US Department of Agriculture must compensate me and my family for crops that could not be planted, because thousands of acres of land were too contaminated with hazardous chemical wastes from Augusta's sewage sludge.[3] Our dairy, which was once one of Georgia's most productive dairy farms, was destroyed by the heavy metals, PCBs, chlordane, and other hazardous wastes that local industries dumped into Augusta's sewer system.

How It Happened

In 1976, Congress enacted the Resource Conservation and Recovery Act (RCRA) for controlling all solid hazardous wastes from "cradle to grave," i.e., from the time that they are created until the time they are destroyed or safely sealed and permanently buried. "Hazardous wastes" include toxic chemicals, radioactive materials, and biological (infectious) wastes that meet certain criteria for being dangerous or potentially harmful to human health or the environment. They can be liquids, solids, contained gases, or sludges.

EPA regulations established under RCRA specifically exclude mixtures of hazardous wastes and domestic sewage passing through publicly owned treatment works (POTW), i.e., sewage treatment plants. To qualify under this exclusion: (1) the materials in the sewer line to which hazardous wastes are added must be domestic sewage; (2) the mixture of hazardous wastes and domestic sewage must flow into a POTW; and (3) any hazardous wastes in excess of thirty-three pounds per month must be "pretreated" before being discharged into sewer lines. Pretreatment standards are designed to protect wastewater treatment plants from nondomestic wastes that may cause explosion or fire, or interfere with the treatment process. They are also aimed at improving the quality of effluents and sludges so that they can be used as fertilizers and soil amendments (biosolids).

In Augusta, the pretreatment program was so lax that it essentially did not exist. Each industrial discharger applied for a pretreatment permit that limited the number of constituents that were monitored in the discharged effluent. Thousands of pounds of chemicals were dumped into the sewers every day that were not monitored at all. Each industrial discharger self-reported the contents of the effluent discharged into the sewer lines. Even if there were gross violations of the pretreatment standards, there was not one instance in the history of Augusta where a discharger of hazardous wastes into the sewer lines was shut down or prevented from discharging into Augusta's sewer system.

Local metal plating operations and manufacturers of pharmaceuticals, artificial sweeteners, and other products dumped their wastes into the sewers of Augusta, Georgia. As toxic chemicals made their way to the wastewater treatment plant, they mixed with the human wastes and concentrated in the sewage sludge in settling tanks.

From there, the sludge was pumped into digesters to reduce levels of disease-causing bacteria and viruses. A small battery of tests developed by EPA was performed to determine the concentrations of nine heavy metals, a few other chemical parameters including nitrogen, and the levels of at least one "indicator" pathogen.

Employees of the Messerly Wastewater Treatment Plant reported their results to the Georgia Environmental Protection Division (EPD), as required under federal and state environmental laws since Augusta's land application program began in 1979. They gave the city's processed sewage sludge a passing grade as "Class B biosolids" and had it trucked out to local farmers, including to our farm and the farm owned by the Boyce family.

Augusta assured us that the city's sewage sludge was completely safe for fertilizing food-chain crops.

The only problem was that Augusta's digesters and other critical equipment were not working properly—sometimes not at all. The pH of the city's "fertilizer" was so low that it dissolved metal fences and parts of the building where lab tests were performed. Employees tested only one of two waste streams of sewage sludge, and those results showed that the sludge that was tested contained hazardous levels of PCBs, chlordane, heavy metals, and other highly toxic wastes.

To appear to be in compliance with the federal Clean Water Act and other environmental laws, city officials routinely altered or outright invented the numbers they reported to the EPD.[4] Records concerning how much sludge was applied per acre were manipulated, and levels of metals in different batches of sludge were averaged to make it appear that annual maximum loading rates for molybdenum and cadmium were not exceeded.

The total amounts of sewage sludge that Augusta applied each year to area farms could not be accurately reconstructed. Different sets of records were kept for amounts of sludge hauled by city and contract employees, and the EPD lost all of Augusta's annual reports showing the combined amounts. The city also lost all of its files showing the amounts of sludge hauled by its contractors. The combined totals reflected in field update reports, the city's only remaining records showing how much sludge was hauled, were inconsistent. Neither EPA, EPD, nor the University of Georgia has ever produced the records EPA and UGA authors used to create summaries of Augusta's historical data, which they published in a scientific journal in 2003.

What is certain is that had Augusta complied with the law, it would have incinerated or buried its sewage sludge as hazardous wastes. Instead, city workers cooked the books to keep from spending the tens of millions of dollars it would have taken to upgrade Augusta's dilapidated wastewater treatment system and produce sewage sludge that could legally be land applied.[5]

The case is not that Augusta lacked the funds to make the needed repairs. The Clean Water Act allows municipalities to collect user fees for upgrading treatment systems to meet federal and state environmental standards. City officials, however, diverted these proceeds to the city's general fund. The few repairs and improvements that were made were covered by low-interest government loans. To qualify for these loans, city officials relied on their false and fabricated environmental monitoring data to certify that the wastewater treatment system complied with the Clean Water Act.

As we and the Boyce families used Augusta's sewage sludge to fertilize forage crops, we noticed that our land was becoming more and more acidic. To continue growing crops, we applied large amounts of lime to raise the pH—first on our farm in 1985 and then on the Boyces' farm in 1996. But as soon as we did, the dairy herds developed an odd reddish tinge to their fading coats, a symptom of molybdenum poisoning. Molybdenum, a toxic heavy metal that attacks the liver and kidneys, dissolves at a very high pH, such as when lime is added. Molybdenum was but one of many toxic chemicals in Augusta's sludge that city officials were either underreporting or not reporting at all.[6]

Milk production from both of our dairies plummeted. Within months, many cows looked emaciated and, on our farm, developed *Salmonella* infections. Many of the cattle on both farms developed various infections and looked as if they were suffering through the last stages of AIDS. Veterinarians and other experts tested soil and forage samples as well as liver and kidney tissue samples. They found high levels of cadmium and other sludge-related contaminants. When the experts finally figured out what was happening, they fed one of the herds forage not grown with sewage sludge. Those animals slowly recovered over a period of two years. In the end, both of our family-owned dairy businesses were destroyed.

In 1998, my family and the Boyce family sued the City of Augusta over damages caused by hazardous wastes in the city's sewage sludge. EPA dispatched Robert Brobst from Region 8 in Denver, Colorado, to investigate. Brobst headed EPA's Biosolids Incident Response Team (BIRT). Brobst had investigated at least one other incident involving cattle, and had ruled that sludge was not the cause.

The EPD responded to our lawsuits by auditing the Messerly Wastewater Treatment Plant in Augusta. They found clear evidence that the city's environmental monitoring reports were being fudged to cover up high levels of contaminants—just as we and our experts discovered Augusta had been doing for decades. One of the reasons Augusta was fabricating data is because the city was not enforcing federal and state pretreatment regulations. The auditors recommended that Augusta's land application program be shut down immediately and the sludge be buried as hazardous wastes.

The Gatekeepers

You are probably thinking by now that this is a story about corrupt city officials being sent to prison. That is what should have happened. Government forms that were used for reporting their false and fabricated environmental monitoring data included a warning in bold-faced type that it is a criminal violation, punishable by fines and imprisonment, to knowingly report false data under the Clean Water Act. This was a clear case of fraud for EPA's criminal investigation division to refer to the US Department of Justice for prosecution.

But that never happened. Test results from soil and forage samples collected from our farm and the Boyce farm indicated that the dairy cows could have died from ingesting levels of molybdenum that are *permissible* under EPA's 503 sludge rule. In other words, what happened on our dairy farms suggested that EPA's sludge rule may have a major loophole—one that allows toxic heavy metals and other pollutants to contaminate food chain crops and milk supplies. Federal bureaucrats EPA's Office of Water, who developed EPA's sludge regulations, had too much to lose if local Augusta officials were held accountable.

EPA headquarters was not unprepared to deal with the bad news coming from Augusta. The ink on our lawsuits had hardly dried when architects of EPA's 503 rule engaged UGA in a strategy for rebottling the evil genie of Augusta, Georgia.[7] Their plan, which Walker initiated in November of 1998, was to get city officials to provide Robert Brobst with a "scientifically reliable" version of Augusta's historical reports showing that sewage sludge spread on either or both of the McElmurray and Boyce farms from the late 1980s through the late 1990s, when the cattle died, was safe for growing forage crops. Prior to 1999, however, the reports were "in shambles" and the data were "sloppy."[8] The reports would require more fudging and inventing.

To get the data into peer-reviewed scientific literature, EPA funded UGA land application specialist Julia Gaskin to publish a research study coauthored by Brobst. Brobst provided Gaskin with Augusta's fabricated data upon which to base the report. Later on, the plan included giving the article to the National Academy of Sciences to use in a 2002 report.[9] If all went well, the research article and academy report would be introduced as evidence at our jury trials.

In 1999, when the Gaskin study was conducted, Allen Saxon was rehired by the city and went to work "fudging" and "inventing" a new and improved version of Augusta's data, which Brobst needed to publish in a scientific journal. Brobst summarized and tabulated Saxon's work product for the Gaskin article. When Saxon and Brobst were finished, years of data that were once in "shambles" now fit tidily into a single table complete with what appeared to be statistically valid means, standard deviations, and maximum pollutant values that could pass muster at almost any reputable scientific journal.[10]

Everything actually worked quite well up until the time US District Court Judge Anthony Alaimo did what no one expected. He spent weeks methodically and meticulously combing through court proceedings and mountains of related testimony and exhibits in our cattle cases, and in Dr. David Lewis's Labor Department case as well, until he pieced the puzzle together. Judge Alaimo ruled that Augusta's reports, which Brobst used in the UGA study and the Department of Agriculture case, were "incomplete," "unreliable," "fudged," "fabricated," and, in some cases, "invented."

Using nitrogen data, which Allen Saxon admitted under oath were off by four orders of magnitude, plus sewage sludge application rates, which Judge Alaimo described as "invented," along with metals concentrations, which Judge Alaimo described as "fudged," Gaskin and her coauthors concluded that Augusta's sewage sludge was applied at agronomic (proper nitrogen) rates and generally met federal and state requirements for levels of regulated metals.

Another member of EPA's BIRT, Robert Bastian, emailed the National Academy of Sciences panel a copy of Gaskin's draft manuscript in 2001.[11] The panel used the manuscript's preliminary, unpublished data to discount our lawsuits and conclude that there was no evidence that sewage sludge applied under EPA's 503 rule has ever harmed public health or the environment. In a national press release issued by UGA when the paper was published in 2003,[12] Julia Gaskin announced:

> *Some individuals have questioned whether the 503 regulations are protective of the public and the environment. This study puts some of those fears to rest.*

Finally, Augusta's attorney, James Ellison, turned on the overhead projector and illuminated the courtroom in Atlanta where my case was under appeal. He displayed the Gaskin article page by page until he came to the conclusion at the end: "Overall, forage quality from fields with long-term application of biosolids was similar to that having only commercial fertilizer and should not pose a risk to animal health." When all was said and done, a jury awarded the Boyces only $550,000 in damages, and our case settled out of court for $1.5 million. These amounts were not even enough to pay our experts, much less make a dent in the tens of millions of dollars that each of our families lost when our dairy farm businesses collapsed.

The Mehan Letter

Brobst was more successful at using his and Gaskin's article to dismiss a public petition on sewage sludge filed with EPA in 2003 by seventy-three farm, health, and environmental organizations. The groups called for a moratorium on land application of sewage sludge until the scientific issues raised by the Boyce verdict and three human deaths linked to sewage sludge could be resolved. EPA assistant administrator G. Tracy Mehan III rejected the petition for the land application moratorium based upon information provided by Bastian and Brobst.

That November, Bastian emailed Madolyn Dominy at EPA-Region IV Atlanta a version of the letter he and Brobst were preparing for Mehan to sign.[13] Bastian wrote:

> *Madolyn. I have been drafted by OST to develop a write-up on the Augusta, GA, case to include in the petition. [The attached version is] such a write-up developed from various materials that... have been provided to me by various sources that incorporates some suggestions that Bob Brobst and I came up with during a conversation earlier today. Do you know if the City of Augusta has or plans to appeal the jury award? Please let me know what you think of this write-up. Bob Bastian*

Assistant Administrator Mehan's final letter issued on Christmas Eve used the Gaskin study to dismiss the jury verdict in favor of the Boyce family.[14] Mehan wrote:

> *On February 2, 1999, Region 4 staff and the BIRT met with University of Georgia veterinarian scientists and soil scientists to discuss the livestock deaths and the University's possible participation in assessing soil and forage characteristics in Burke and Richmond Counties. On August 5, 1999, EPA Headquarters issued a grant to the University of Georgia.... This effort resulted in the publication of a paper entitled Long-Term Biosolids Application Effects on Metal Concentrations in Soil and Bermudagrass Forage (Gaskin et al., 2003).*
>
> *The University of Georgia's findings of their analyses of trace metals levels in soils and feed that were implicated in the*

Georgia case. The paper indicates "that toxic levels of metals have not accumulated in the soils due to long-term biosolids application. Overall forage quality from the biosolids-amended fields was similar to that of commercially fertilized fields."

Thus, EPA's investigation of the site and the sewage sludge did not find any substantiation to the allegations that exposure to sewage sludge applied to the pasture land caused illness or death of the dairy cattle that grazed on the pasture.

According to John Walker's typewritten notes of a telephone conversation he had with Dominy in November of 1998, Dominy told Walker that analyses of soil samples from our farms showed that the land was contaminated with 30 ppm (mg/kg) of molybdenum compared with a background concentration of only 0.5 ppm. However, the Gaskin study of other farms reported mean soil molybdenum concentrations of only 0.089 (± 0.041) ppm (Table 3, p. 149). EPA officials involved in drafting the Mehan letter knew that the results in the Gaskin paper grossly misrepresented contaminant levels found on our farm and the Boyce farm.

Ruling by Judge Anthony A. Alaimo

In January of 2003, we filed for economic relief under the federal Farm Bill to cover losses from corn and cotton crops during the 1998–2001 growing seasons. We claimed that our land was too polluted by PCBs, chlordane, heavy metals, and other hazardous wastes in Augusta's sewage sludge to grow food-chain crops. USDA, however, rejected these claims based mainly on information in the Gaskin study supplied by the study's EPA coauthor, Robert Brobst.

In February of 2008, Judge Anthony Alaimo ruled that the USDA's conclusions were "arbitrary and capricious." Regarding Brobst's summaries of Augusta's historical data concerning sludge application rates and pollutant concentrations, Alaimo wrote, "Although there is a broad consensus that Augusta's reports were unreliable, incomplete, and in some cases, fudged, the City's information is an integral part of this case."

To support these findings, Judge Alaimo referred to detailed analyses of Augusta's reports performed by our experts, an audit in which the EPD confirmed the conclusions made by these experts, and other key evidence such as sworn testimony taken from employees working for the City of Augsuta. For example, Judge Alaimo found:

> *There is also evidence that the City fabricated data from its computer records in an attempt to distort its past sewage sludge applications.... In January 1999, the City rehired [former City of Augusta supervisor Allen] Saxon to create a record of sludge applications that did not exist previously.*

In addition to ruling that environmental data summarized in the Gaskin paper were fudged, fabricated, and invented by the City of Augusta, Judge Alaimo ruled that EPA and the USDA relied on data collected in 1999 (when the Gaskin study was performed) while ignoring ample data collected as much as a decade earlier, at or about the time our cattle were dying. These data proved that our property was highly contaminated. Judge Alaimo wrote:

> *Other specific evidence showed that heavy metals were found at levels that were above the regulatory limits on the McElmurrays' farm, making the land unfit for food grown for human consumption. On one piece of property alone, antimony levels registered at 96.8 ppm, while the regulatory limit was 4 ppm. Arsenic registered at 44.2 ppm, more than twice the amount allowed by law. Cadmium was found at a level of 6.41 ppm, which was more than three times the level deemed safe under the law. Selenium registered at 5.4 ppm, although the cleanup standard provided under the law was set at 2 ppm. Thallium was found at 51.6 ppm on that particular piece of property, although the regulatory limit is 2 ppm.*

How Widespread Are the Problems?

My attorney, Mr. Hallman, invited Dr. David Lewis to meet with the experts working on our cases in April of 2003. This is the first time I ever met Dr. Lewis. We were surprised to learn that Dr. Lewis and our veterinarians and other experts had independently come to the same conclusion regarding infections linked to sewage sludge. Dr. Lewis and the scientists working with him concluded that many people living near land application sites, who breathed sewage sludge dusts blowing from the fields, suffered from chemical irritation of the skin, eyes, and respiratory tract. This chemical irritation, Dr. Lewis postulated, lead to a variety of infections.

Our experts had concluded that chemical wastes in Augusta's sewage sludge sickened and killed our cattle in the same way, by attacking internal organs when the contaminated forage was eaten. Once the organs were damaged, the animals started contracting various kinds of infections. My father and I both experienced the same symptoms described in Dr. Lewis's research articles. We stayed on antibiotics. Then, as my father's condition worsened, he had to be kept on massive doses of corticosteroids. He almost died and still suffers serious medical problems from having worked in the sludge-amended fields and from getting steroid treatments. We never made the connection between our illnesses and what was happening to our dairy herds—not until we read the research articles published by Dr. Lewis.

Dr. Lewis provided us with many of the documents that he had collected when he worked on sewage sludge at EPA. These documents filled in many of the gaps in what we knew about what was happening in Augusta. We learned, for example, that EPA set up a cooperative agreement with the Water Environment Federation in 1992 to promote sewage sludge as safe and beneficial. The agreement included studying (and no doubt dismissing) ten "unsubstantiated horror stories." One internal EPA memo discussed the problems on our two dairy farms, mine and the Boyces'. The memo stated: "Biosolids Horror Stories. We asked Bob [Brobst] for real life examples of adverse environmental effects from biosolids. Bob sent us a list of sites with groundwater contamination."

The tables of field data attached to the memo indicated widespread groundwater contamination with nitrates and heavy metals at multiple sites in a study conducted in California, Colorado, Georgia, Illinois, Maine, Minnesota, New Mexico, Nebraska, and South Carolina. I do not believe that Augusta is unique. We have heard from dairy farmers elsewhere in Georgia, and in other states as well, where cattle were sickening and dying after being fed forage crops fertilized with sewage sludge. In one case, autopsies demonstrated that molybdenum poisoning was the likely cause of death.

We also learned from Dr. Lewis's documents that, in 1992, EPA's Office of Research and Development (ORD) identified six major weaknesses in the science used to support the 503 sludge rule. According to an Inspector General report ten years later, EPA's Office of Water never funded ORD to fix any of these problems.[15] OW claimed that it did not fund ORD because research on sewage sludge became a low priority in 1993 under the Clinton administration. OW, however, worked with the WEF from 1992 to 1999 to put tens of millions of dollars in congressional earmarks into funding proponents of land application to publish research supporting the 503 rule. This research did not find any problems with sewage sludge—only benefits.

Some of the weaknesses that ORD identified were the very problems that showed up on our dairy farm and on the Boyce farm as well. For example, ORD wanted to determine the bioavailability of sewage sludge contaminants for uptake by plants and animals. Our cattle were killed when they ingested sludge contaminants taken up by plants. This is also how milk on the Boyce farm became contaminated.

The ORD found weaknesses in the science EPA uses to support land application of sewage sludge, which have existed since the program first began. For example, the ORD pointed out that we need to understand long-term changes at land application sites, including changes in soil pH, land use, and the capacity for sewage sludge to bind chemical contaminants. Again, these kinds of changes are exactly what led to our cattle being poisoned. Our soil pH gradually had dropped over years of sludge applications. Then, when we switched to growing alfalfa—a change in land-use—we had to add lime. The lime caused the soil to lose its ability to bind molybdenum, which had built up to high levels from Augusta's sewage sludge. If the ORD had been able to address the weaknesses its scientist had identified in the sludge rule, and the Office of Water had fixed these problems, then the Boyce family and my family would not have lost our dairy businesses.

Before Dr. Lewis stopped doing the research, UGA approved a grant proposal that he submitted for a Swiss foundation to fund his research. Our farm was going to participate in the study, in which we planned to collect and analyze soil and groundwater samples. We also planned to collect milk samples from dairies using sewage sludge and test them for heavy metals and priority pollutants. This project would have addressed some of the weaknesses the ORD had identified. But, once again, senior EPA officials in the Office of Water stopped the work from being done.[16]

Conclusion

In conclusion, ORD clearly identified many of the main weaknesses with the 503 sludge rule when it was first reviewed in 1992. Office of Water has prevented ORD from addressing any of these weaknesses for the past sixteen years and tried to cover up any harm to public health or the environment.

The same few people have run this program since the 1970s, and the program has only gotten more inept and corrupt with every passing year.

The first step toward fixing problems with land application of sewage sludge, therefore, is to clean up the longstanding corruption associated with this program in EPA's Office of Water, take the millions of dollars the Office of Water is funneling to it supporters with congressional earmarks, and redirect all future funding in this area to ORD.

APPENDIX VII: THE NUREMBERG CODE

1. The voluntary consent of the human subject is absolutely essential.

    This means that the person involved should have legal capacity to give consent; should be so situated as to be able to exercise free power of choice, without the intervention of any element of force, fraud, deceit, duress, over-reaching, or other ulterior form of constraint or coercion; and should have sufficient knowledge and comprehension of the elements of the subject matter involved, as to enable him to make an understanding and enlightened decision. This latter element requires that, before the acceptance of an affirmative decision by the experimental subject, there should be made known to him the nature, duration, and purpose of the experiment; the method and means by which it is to be conducted; all inconveniences and hazards reasonably to be expected; and the effects upon his health or person, which may possibly come from his participation in the experiment.

    The duty and responsibility for ascertaining the quality of the consent rests upon each individual who initiates, directs , or engages in the experiment. It is a personal duty and responsibility which may not be delegated to another with impunity.

2. The experiment should be such as to yield fruitful results for the good of society, unprocurable by other methods or means of study, and not random and unnecessary in nature.

3. The experiment should be so designed and based on the results of animal experimentation and a knowledge of the natural history of the disease or other problem under study, that the anticipated results will justify the performance of the experiment.

4. The experiment should be so conducted as to avoid all unnecessary physical and mental suffering and injury.

5. No experiment should be conducted, where there is an a priori reason to believe that death or disabling injury will occur; except, perhaps, in those experiments where the experimental physicians also serve as subjects.

6. The degree of risk to be taken should never exceed that determined by the humanitarian importance of the problem to be solved by the experiment.

7. Proper preparations should be made and adequate facilities provided to protect the experimental subject against even remote possibilities of injury, disability, or death.

8. The experiment should be conducted only by scientifically qualified persons. The highest degree of skill and care should be required through all stages of the experiment of those who conduct or engage in the experiment.

9. During the course of the experiment, the human subject should be at liberty to bring the experiment to an end, if he has reached the physical or mental state where continuation of the experiment seemed to him to be impossible.

10. During the course of the experiment, the scientist in charge must be prepared to terminate the experiment at any stage, if he has probable cause to believe, in the exercise of the good faith, superior skill, and careful judgment required of him, that a continuation of the experiment is likely to result in injury, disability, or death to the experimental subject.

Source: "Trials of War Criminals before the Nuremberg Military Tribunals under Control Council Law No. 10," Vol. 2, pp. 181–182. Washington, DC: US Government Printing Office, 1949. www.hhs.gov/ohrp/archive/nurcode.html.

CHAPTER NOTES

Foreword

1. Ibsen M. *An Enemy of the People* (1882).
2. In a story told by Norwegian playwright Henrik Ibsen, a medical officer, Dr. Thomas Stockmann, wants to publish information about his town's water baths being polluted with disease-causing microorganisms and chromium from the local tannery. The baths, which were reputed to have health benefits, were a major source of the town's income. The mayor, who is Stockmann's brother and supervisor, casts Stockmann as the public enemy, as do all of the townspeople, including Stockmann's closest friends. Dr. Stockmann, as a result, loses his government position and private medical practice. (Provided by David Lewis)
3. Edwards M. AEESP. Saving mankind from itself for 40 years (and counting). Presidents Letter January 14, 2004. Accessed 12/3/2013 at https://www.aeesp.org/sites/default/files/publications/AEESPNL.39.1.2004.pdf
4. Hiltzik M. Science has lost its way, at a big cost to humanity. October 27, 2013. *Los Angeles Times*.
5. *The Economist*. Unreliable Research. Trouble at the Lab. Scientists like to think of science as self-correcting. To an alarming degree, it is not. October 3, 2013.
6. *The Economist*. Problems with scientific research. How science goes wrong. Scientific research has changed the world. Now it needs to change itself. October 19, 2013.
7. Broad W, N Wade. Betrayers of the Truth. Fraud and Deceit in the Halls of Science. Century Publishing Co. Ltd, London (1983).
8. Whyte WH. The Organization Man. Simon and Schuster, New York (1956).
9. Lewis D. Personal Communication, Dec. 2, 2013.

10. Lewis D. How to study research ethics? A dialogue between different research perspectives. Littauer Panel III – *Enacting Ethics: Conflicts of interest in practice*. Chair: Ellen Bales, Harvard Kennedy School, STS Program. May 6, 2011.

11. Holder E. (2004). Summary of Investigation Reported to the Board of Directors of the District of Columbia Water and Sewer Authority. http://www.washingtonpost.com/wp-srv/metro/specials/water/wasa071604.pdf (accessed Aug. 2, 2004).

12. House Government Reform Committee. Public Confidence, *Down the Drain: The Federal Role in Ensuring Safe Drinking Water in the District of Columbia*. Hearing March 5, 2004.

13. Edwards M. Testimony to the 108th Congress of the United States. *Lead in DC Drinking Water*. House Committee on Government Reform. March 5, 2004. 29 pages.

14. Fisheries, Wildlife and Water Subcommittee of Environment and Public Works. *Oversight of Drinking Water in the District of Columbia*. Hearing April 7, 2004.

15. House Government Reform. *DC Lead Crisis*. Hearing May 21, 2004.

16. Energy and Commerce Committee Subcommittee on Environment and Hazardous Materials U.S. House of Representatives. *Lead in DC Water*. Hearing. July 22, 2004.

17. House Government Reform. Oversight of DC Lead in Water. Hearing. March 11, 2005.

18. Government Accountability Office (2006). Report to Congressional Requesters. EPA Should Strengthen Ongoing Efforts to Ensure That Consumers Are Protected from Lead Contamination. GAO-06-148

19. Blood lead levels in residents of homes with elevated lead in tap water–District of Columbia, 2004. CDC *Morb. Mort. Weekly Rep.* 2004, 53, 268–270.

20. Edwards M. Discussion: Effect of Changing Water Quality on Galvanic Coupling. *J.AWWA* 104(12) 65-82 (2012).

21. Guidotti, TL; Calhoun T.; John O. Davies-Cole, J.O.; Knuckles, M.E.; Stokes L.; Glymph, C; Lum G; Moses, M.S.; Goldsmith, D.F.; Ragain, L. Elevated lead in drinking water in Washington DC, 2003-2004: The public health response. *Environ. Health Perspect.* 2007, 115, 695-701.

22. A Public Health Tragedy: How Flawed CDC Data and Faulty Assumptions Endangered Children's Health. May 20, 2010. US

Congressional report by the Oversight Committee on Science and Technology.

23. Edwards M. Written Testimony to the House Committee on Science and Technology. May 20, 2010.

24. CDC. Notice to Readers. Examining the Effect of Previously Missing Blood Lead Surveillance Data on Results Reported in MMWR. *Morbidity and Mortality Weekly Report.* 59(19); 592. May 21, 2010.

25. Frieden TR. Lead in DC Water: Still more to do. *Washington Post.* June 27, 2010.

26. Virginia Tech professor uncovered truth about lead in D.C. water. *By Robert McCartney* Sunday, May 23, 2010.

27. Edwards M. and 21 other individuals and organizations. Personal communication to Katherine Sebellius. False Statement in the May 21, 2010 CDC *Morbidity and Mortality Weekly Report.* Sent May 27, 2010.

28. CDC's Botched Handling of 2004 District Lead Scare Reveals Toxic Bureaucracy. *By Robert McCartney* Sunday, May 27, 2010.

29. SSIT Announces IEEE Barus Award to Environmental Engineer Marc Edwards. *Technology and Society Magazine, IEEE* 31(2) 2012.

30. Rushton JP. Victim of Scientific Hoax (Cyril Burt and the Genetic IQ Controversy). *Society,* 31, 40-44 (1994).

31. Rushton JP. (2002). New Evidence on Sir Cyril Burt: His 1964 Speech to the Association of Educational Psychologists, *Intelligence,* 30, 555-567 (2002)

32. Hunt E. Human Intelligence. Cambridge University Press, New York (2011).

33. *Ibid,* Ibsen (1882).

34. Heroic By Nature, Cowardly by Convenience. TEDx Talk Virginia Tech, Beyond Boundaries. November 9, 2013. http://www.youtube.com/watch?v=apZynV7lfic

Prologue

1. Kohn SM. (2011) Conclusion: Whistleblowing and The American Dream, The Whistleblowers Handbook, pp. 313-4. Lyons Press, Guilford, CT.

2. Union of Concerned Scientists, Cambridge, MA. Hundreds of EPA scientists report political interference over last five years. Apr. 23, 2008.
3. Personal communication. Meeting between Lewis D, Cohen B and Cooke CE, Special Assistant to Congressman Ralph Hall (1999).
4. Chapter 5, section titled: "UGA and Republicans Torpedo Senate Hearings."

President Dwight D. Eisenhower's Farewell Address

1. Dwight D. Eisenhower Presidential Library. The Farewell Address, January 17, 1961, http://www.eisenhower.archives.gov/research/online_documents/farewell_address/1961_01_17_Press_Release.pdf

Chapter 1

1. CDC. Recommended infection-control practices for dentistry. 1993. *Morbid. Mortal. Wky. Rep.* 42:1-12.
2. *Lancet*, Risk of HIV transmission during dental treatment. 1992;340:1259 [Editorial].
3. HHS, Center for Devices and Radiological Health, J. S. Benson, Director. Dear Doctor letter, Dental Handpiece Sterilization. Sept. 28, 1992.
4. Breo DL, The dental AIDS cases—Murder or an unsolvable mystery? *JAMA* 1993; 270, 2732-4.
5. Lewis DL, Arens M, Appleton S, Nakashima K, Ryu J, Boe RK, Patrick J, Watanabe D, Suzuki M. Cross-contamination potential with dental equipment. *Lancet* 1992; 340:1252-4; Lewis DL, Arens M. Resistance of microorganisms to disinfection in dental and medical devices. *Nature Med.* 1995;1:956-8.
6. CDC. Possible Transmission of Human Immunodeficiency Virus to a Patient during an Invasive Dental Procedure. *MMWR* 39: 489 (July 27, 1990).
7. Lambert B. Kimberly Bergalis Is Dead at 23; Symbol of Debate Over AIDS Tests. NY Times, Dec. 9, 1991.
8. Cheek M, Drill suspect in AIDS case. *The Stuart News*, July 11, 1991.
9. Lewis DL, Arens M., *Nature Med.* 1995. Note: We did not identify Dr. Acer's patient as the source of any of the blood samples used in

the study. HIV entrapped in dental and endoscope lubricants survived disinfection for at least forty-eight hours, but less than two weeks.

10. Gooch B, Marianos D, Ciesielski C, Dumbaugh R, Lasch A, Jaffe H, Bond W, Lockwood S, Cleveland J, Lack of evidence for patient-to-patient transmission of HIV in a dental practice. *J. Am. Dent. Assoc.* 1993;124, 38-44.

11. *Ibid*, Lewis DL, Arens M., *Nature Med.* 1995.

12. Lewis DL, Boe RK. 1992. Cross-infection risks associated with current procedures for using high-speed dental handpieces. *J. Clin. Microbiol.* 30:401-406.

13. Lewis DL, Arens M, Appleton S, Nakashima K, Ryu J, Boe RK, Patrick J, Watanabe D, Suzuki M. Lancet 1992; 340:1252-4.

14. *James Sharpe and Jeanne Sharpe v. Anthony E. Breglio, DMD and Robert A. Breglio, DMD.* Superior Court Department of the Trial Court of the Commonwealth of Massachusetts, 1993. CA-93-034. 20.

15. *Sharpe et al. v. Breglio et al.*, Defendant Anthony E. Breglio, Answers to Second Set of Interrogatories, CA-93-034. 20, Sept. 7, 1994.

16. Bollinger RC, Brookmeyer RS, Mehendale SM, Paranjape RS, Shepherd ME, Gadkari DA, Quinn TC. Risk factors and clinical presentation of acute primary HIV infection in India. *J. Am. Dent. Assoc.* 1997;278(23):2085–2089.

17. Lewis D to Zabin A, July 5, 1995 [Letter].

18. Roche BJ. Case contends AIDS infection by dental tools. Boston Globe Feb. 2, 1996.

19. Hilkevitch J. It's A Mystery Why Girl, 11, Has AIDS, *Chicago Tribune*, June 20,1993; Mystery of AIDS Girl Has a Twist, *Chicago Tribune*, July 18, 1993.

20. *Ibid.*

21. U.S. Department of Labor, Bureau of Statistics. 2010.

22. Centers for Disease Control and Prevention. *MMWR.* 35, 237 (1986).

23. Christensen GJ. Infection control: some significant loopholes. *J. Am. Dent. Assoc.* 1991; 122:99-100.

24. Verrusio AC, Neidle EA, Nash KD, Silverman S Jr, Horowitz AM, Wagner KS. The dentist and infectious diseases: a national survey of attitudes and behavior. *J Am Dent Assoc.* 1989 May;118(5):553-62. Erratum in: *J Am Dent Assoc* 1989 Jun;118(6):684.

25. Alter MJ, Kruszon-Moran D, Nainan OV, McQuillan GM, Gao F, Moyer LA, Kaslow RA, Margolis HS. The prevalence of hepatitis C

virus infection in the United States, 1988 through 1994, *N Engl J Med.* 1999 Aug 19;341(8):556-62.

26. Centers for Disease Control and Prevention. A comprehensive immunization strategy to eliminate transmission of hepatitis B virus infection in the United States: recommendations of the Advisory Committee on Immunization Practices (ACIP); Part 1: Immunization of Infants, Children, and Adolescents. *MMWR* 2005;54(No. RR-16):[inclusive page numbers].

27. Centers for Disease Control and Prevention, 1998. http://hepnet.com/hepc/cdc98/intro.html.

28. Kuehn R. 2004. Suppression of Environmental Science, *American Journal of Law & Medicine*, 30:333-69

Chapter 2

1. This discussion is drawn, in part, from an article I published in the *International Review of Modern Surgery* (1999, pp. 77-83) in association with the International Federation of Surgical Colleges as part of his official duties for the U.S. Environmental Protection Agency.

2. CDC, Stopping *C. difficile* Infections, CDC Vital Signs, March 2012. http://www.cdc.gov/VitalSigns/Hai/StoppingCdifficile/

3. CDC, Healthcare Infection Control Practices Advisory Committee (HICPAC), Guideline for Disinfection and Sterilization in Healthcare Facilities, 2008. http://www.cdc.gov/hicpac/Disinfection_Sterilization/13_0Sterilization.html

4. Spaulding EH. 1968. Chemical disinfection of medical and surgical materials. In. Lawrence, C.A. and S.S. Block, eds. *Disinfection, sterilization, and preservation*. Philadelphia: Lea & Febiger.

5. *Ibid*, see Ref. 3, Crow, Malchesky et al., Seballos et al., 1995.

6. Lewis DL, Arens M. Resistance of microorganisms to disinfection in dental and medical devices. *Nature Med.* 1995;1:956-8.

7. Bond WW, Favero MS, Mackal DC, Malison GF. Sterilization or disinfection of flexible fiberoptic endoscopes. *AORN J.* 1979; 30:350-352; Welter IVD., Williams CB, Jeffries DJ, *et al.* Cleaning and disinfection of equipment for gastrointestinal flexible endoscopy: interim recommendations of a Working Part of the British Society

for Gastroenterology. *Gut.* 1988; 29:1134-51; Cowan RE, Manning AP, *et al.* Aldehyde disinfectants and health in endoscopy units. 1993; *Gut.* 34:1641-5.

8. Olympus Optical Co., Ltd. Tokyo, Japan. 1995. U.S. Patents 5,408,991; 5,419,311; 5,431,150; 5,458,132; 5,458,133.

9. Lewis DL, Arens M. Microbially contaminated medical/dental device lubricants. Biomedicine '96, Washington, DC, May 1996.

10. Lewis DL, Arens M. Resistance of microorganisms to disinfection in dental and medical devices. *Nature Medicine* 1995; 1:956-958.

11. Jensen MN. Bacteria may hide in hunks of gunk. *Science News* 1998;153:137.

12. Cheung RJ, Ortiz D, DiMarino AJ. GI endoscopic reprocessing practices in the United States. *Gastrointest. Endosc.*, 1999, 3: 362-64.

13. Johnson & Johnson Medical, 1995; Michele TM, Cronin WA, Graham NMH, Dwyer DM, Pope DS, Harrington SH, Chaisson RE, Bishai WR. 1997. Transmission of Mycobacterium tuberculosis by a fiberopitc bronchoscope. Identification by DNA fingerprinting. JAMA. 278:1093-1095.

14. Tucker RC, Lestini BS, Marchant RE. Surface analysis of clinically used PTFE endoscopic tubing treated by the STERIS PROCESS. *ASAIO Journal* 1996;42:306-313.

15. Zuber TJ. Disposable sheath endoscopes: a new office technology. American Family Physician. 1994;50: 1465-68; Schroy PC, Wilson S, Afdhal N. Feasibility of high-volume screening sigmoidoscopy using a flexible fiberoptic endoscope and a disposable sheath system. Am. J. Gastro. 1996;91: 1331-37; Sardinha TC, Wexner SD, Gilliland J, Daniel N, Kroll M, Lee E, Wexler J, Hudzinski D, Glass D. Efficiency and productivity of a sheathed fiberoptic sigmoidoscope compared with a conventional sigmoidoscope. *Dis Colon Rectum* 1997;40:1248-1253.

16. Olympus Optical Co., Ltd. Tokyo, Japan. U.S. Patents 5,458,132 (1995); 5,536,235 (1996); 5,554,098 (1996); 5,674,182 (1997).

17. Finch S. The dirt on disinfectants. Hippocrates. February 1998, p. 40-47.

18. Wallace, CG, Demicco DD, Agee PM. 1990. Nosocominal pseudoinfection associated with endoscopy processor disinfection using 2% glutaraldehyde. Abstract at Third International Conference on

Nosocominal Infections. Atlanta, GA, July-August 1990; Wallace CG, Agee PM, Demicco DD. Liquid chemical sterilization using peracetic acid. *ASAIO Journal*, 1995;41:151-154.

19. Johnson & Johnson Medical. How the right disinfectant can prevent the wrong diagnosis. Technical advisory JJM, 1995;346 April 1995.

20. Foss D, Monagan D. A National survey of physicians' and nurses' attitudes toward endoscope cleaning and the potential for cross-infection. *Society for Gastroenterology Nurses and Associates*. October, 1992. pp. 59-65.

21. Johnson & Johnson Medical, 1995; Michele TM, Cronin WA, Graham NMH, Dwyer DM, Pope DS, Harrington SH, Chaisson RE, Bishai WR. 1997. Transmission of *Mycobacterium tuberculosis* by a fiberopitc bronchoscope. Identification by DNA fin-gerprinting. *JAMA*. 278:1093-1095.

22. Lewis DL. Lack of HIV transmission in a dental practice. Ann. Intern. Med. 1995;122, 960.

23. U.S. Centers for Disease Control & Prevention. Recommended infection-control practices for dentistry. *Morbid. Mortal. Wky. Rep.* 1993;42:1-12.

24. See Ref. 21, Foss, Monagan, 1992.

25. *Ibid*. Lewis DL, Arens M., Nature Med. 1995.

26. S Finch. The Dirt on Disinfectants. And Editor's Note: The Trouble with Endoscopes; *Hippocrates*, Feb. 1998, p. 3; 40-47; R Davis, Medicine's dirty little secret, *USA Today*, Feb. 18, 1999; A Underwood, Do Scopes Spread Sickness, *Newsweek*, Mar. 1, 1999; J Appleby, Medical community debates scope-cleaning procedures, *USA Today*, Apr. 28, 2002.

27. ABC News, Dirty Scopes, *Good Morning America*, Apr. 18, 2002; PBS *Healthweek*, Endoscopes, Sep. 17, 2000.

28. GW Meyer, An Introduction to Scopes and Sterilization; Lewis DL, A Sterilization Standard for Endoscopes and Other Difficult to Clean Medical Devices, *Pract. Gastoenterol*. 1999; 23:Introduction, 28, 30, 35-6, 42, 51-2, 54, 56.

29. Muscarella LF, Custom Ultrasonics, Inc. Ivyland, PA. *The Q-Net Monthly*, May-Jun., Jul.-Aug., 2003.

Chapter 3

1. North East Biosolids & Residuals Association (NEBRA), The Science of Biosolids Recycling. http://www.nebiosolids.org/index.php?page=science

2. Lewis DL, W Garrison, Wommack KE, Whittemore A, Steudler, Melillo J. 1999. Influence of environmental changes on degradation of chiral pollutants in soils. *Nature* 401:898-901.

3. Bienkowski B. Fish on Prozac: Anxious, anti-social, aggressive, *Environ. Health News,* June 12, 2013.

4. *Ibid.*

5. European Parliament Directorate-General for Research, Directorate A, Division for Industry, Research, Energy & Environment and STOA (Scientific and Technical Options Assessment Unit). Neurotoxicity of Environmental Pollutants. EP/IV/A/STOA/2000/09/04.

6. Lewis DL, Gattie DK, Novak ME, Sanchez S, Pumphrey C. Interactions of pathogens and irritant chemicals in land-applied sewage sludges (biosolids). *BMC Public Health* 2:11 (28 Jun). www.biomedcentral.com/1471-2458/2/11.

7. Marshall J. National Press Club, Washington, DC. Mar. 23, 1999.

8. Stuck in the mud—The Environmental Protection Agency must gather data on the toxicity of spreading sewage sludge [Editorial]; Tollefson J. Raking through sludge exposes a stink. *Nature,* 2008, Vol. 453, p. 258, 262-3, May 15, 2008.

9. Khuder S, Milz SA, Bisesi M, Vincent R, McNulty W, Czajkowski K. Health survey of residents living near farm fields permitted to receive biosolids. *Arch. Environ. Occup. Health* 62, 5–11 (2007).

10. Paris DP, Lewis DL. Chemical and Microbial Degradation of Ten Selected Pesticides in Aquatic Systems. 1973; *Res. Rev.* 45: 95-124.

11. Kamler J, Soria JA. Supercritical Water Gasification of Municipal Sludge: A Novel Approach to Waste Treatment and Energy Recovery. In: *Gasification for Practical Applications,* Yun Y, Ed. ISBN 978-953-51-0818-4. Published: Oct. 24, 2012. DOI: 10.5772/51048.

12. Schmidt CW. The Endocrine Society Issues Statement of Principles. *Environ Health Perspectives.*2012;120(9). See also Fagin D. Toxicology: The learning curve. News Feature. *Nature* 490, 24 Oct. 2012.

13. USDA Pesticide Data Program Report USDA (2010). www. ams. usda.gov/pdp. U.S. EPA. Targeted National Sewage Sludge Survey (TNSSS) Sampling and Analysis Technical Report. EPA-822-R-08-016. Jan. 2009.

14. NEBRA, http://www.nebiosolids.org/index.php?page=faqs

15. *Ibid*, EPA TNSSS.

16. Landrigan P, Lambertini L, Birnbaum L. A Research Strategy to Discover the Environmental Causes of Autism and Neurodevelopmental Disabilities. *Environ Health Perspect*. 2012 July; 120(7): a258–a260.

17. EPA-WEF *National Biosolids Public Acceptance Campaign*, Grant No. CX-820725-01-0, Decision Memorandum. Quigley MJ to Cook MB, July 28, 1992.

18. EPA Cooperative Agreement CR-820725-01-01, Decision Memorandum. Lee RE, Chief, Municipal Technology Branch to Quigley MJ, Director, Municipal Support Division, Sept. 23, 1996.

19. Grey AC, Deputy Executive Director, WEF, to Walker J, EPA, Jun. 12, 1995. Proposal to Amend and Expand Cooperative Agreement, CX-820725-01-3, National Biosolids Public Acceptance Campaign [Cover Letter, Proposal].

20. *Ibid*.

21. *Ibid*.

22. EPA-WEF FY94 Amendment. *National Biosolids Public Acceptance Campaign*, Decision Memorandum. Quigley to Cook, Jul. 30, 1994.

23. *Ibid*, Grey, 1995.

24. *Ibid*, Lee, 1996.

25. *Ibid*.

26. US EPA Report: EPA-600/S1-81-026, 232 p. (Apr. 1981). Sewage Sludge—Viral and Pathogenic Agents in Soil-Plant-Animal Systems. G.T. Edds and J.M. Davidson, Institute of Food and Agricultural Systems, University of Florida. An EPA Project Summary is available online at http://nepis.epa.gov/ by searching 600S181026 or key words in the title of the report. To obtain a copy of the full report, contact the National Technical Information Service, Springfield, VA 22161 (Tele. 703-487-4650). Request Order No. PB 81 179 103.

27. Efroymson, RA, Sample BE, Luxmoore RJ, Tharp ML, Barnthouse LW. Final Report: Evaluation of Ecological Risks Associated with

Land Application of Municipal Sewage Sludge. Oak Ridge National Laboratory. ORNL/TM-13703. Sept. 30, 1998.

28. *Ibid*, p. 197.

29. *Ibid*, Quigley MJ to Cook MB, Jul 30, 1994.

30. EPA, Note of conversation with Bob Brobst, Biosolids Management and Enforcement, EIHWF8-11-0027. May 11, 1999 [Memorandum].

31. McElmurray A, *Lynchburg News & Advance*, Letters to the editor, Sept. 1, 2005.

32. Perciasepe R, EPA Assistant Administrator for Office of Water, to SA Herman, EPA Assistant Administrator for Office of Enforcement and Compliance Assurance. Request for Additional OECA Resources for the Biosolids Program. Apr. 29, 1998 [Memorandum].

33. *McElmurray v. United States Department of Agriculture*, United States District Court, Southern District of Georgia. Case No. CV105-159. Order issued Feb. 25, 2008.

34. Gaskin J, Brobst R., Miller W, Tollner W. Long-term biosolids application effects on metal concentrations in soil and bermudagrass forage. J. Environ. Qual. 32, 146-152 (2003), p. 151.

35. Holmes C, University of Georgia. Sludge study relieves environmental fears. Jan. 29, 2003.

36. National Research Council. *Biosolids Applied to Land: Advancing Standards and Practice*, p. 4. National Academy Press. Washington, DC, 2002.

37. *Lewis v.* EPA, U.S. Department of Labor, Office of Administrative Law Judges, Case No. 97-CAA-7. In retaliation over my 1996 Nature commentary, which discussed gaps in the science EPA uses to support a number of regulations, the DOL found: "Dr. David Lewis was discriminated against by EPA's inquiry into ethics violations at the highest levels and communicating these allegations to members of Congress." Settlement: EPA paid $115,000 ($75,000 in attorneys' fees, $40,000 to plaintiff), and issued a letter stating that I did not violate the Hatch Act or any ethics rules as EPA had claimed; R Russo v. EPA, U.S. Department of Labor, OSHA Case No. 1183524. Oct 2, 2000. OSHA found that EPA assistant administrator Norine Noonan lowered Dr. Russo's performance rating, denied her a Special Act Award and a 5 percent pay bonus, and removed her from her position as laboratory director in retaliation for her approval of my

research article published by Nature in 1999, which raised concerns about EPA's 503 sludge rule. EPA restored Dr. Russo to her position and compensated her for lost wages.

38. *Ibid, McElmurray v. USDA*, 2008.

39. *United States of America, ex rel. David L. Lewis, Ph.D., et al. v. John Walker, Ph.D., et al.*, US District Court, Middle District of Georgia, Athens Division, Case No. 3:06-CV-16, Deposition of Regina Smith, Ph.D., April 27, 2009, p. 73, 81-82.

40. U.S. Environmental Protection Agency. Alexis Strauss, Director, Water Division, EPA Region IX, San Francisco, to Frank J. Doyle, P.E., Director, Department of Environmental Services, Honolulu, HI. December 2, 2003 [Letter].

41. EPA's Office of Inspector General ("OIG") investigated EPA's interactions with Synagro, and Dr. Walker's distribution of Defendant Synagro's white paper. During the interviews, a supervisory OIG agent asked Dr. Walker: "Do you see that you appear to be in cahoots with Synagro?" Dr. Walker replied: "We are not in cahoots with Synagro ... I see we do have an appearance problem." Vanderhoef JJ, Project Manager, Headquarter Audit Division. EPA OIG, Hotline Case No. 2001-32, Oct. 12, 2001 [Memorandum].

42. *Lewis v. EPA*, Office of Administrative Law Judges, Case Nos. 2003-CAA-6, 2003-CAA-5; ARB Case No. 04-117; U.S. Court of Appeals, The Eleventh Circuit, Case No.08-12114HH. The DOL found that Dr. John Walker retaliated against Dr. Lewis in violation of whistleblower protections in federal environmental statutes by distributing Defendant Synagro's white paper containing false allegations of research and ethical misconduct. The DOL determined, however, that EPA was not liable for Dr. Walker's actions because they were not approved by EPA, and because EPA promptly prevented any further distribution of Synagro's false allegations. In 2011, the Eleventh Circuit Court of Appeals affirmed the DOL's decision that EPA was not liable for Dr. Walker's distribution of Synagro's white paper.

43. United States of America, ex rel. David L. Lewis, Ph.D., et al. v. John Walker, Ph.D., et al., US District Court, Middle District of Georgia, Athens Division, Case No. 3:06-CV-16, Deposition of UGA Research Integrity Officer Dr. Regina Smith, April 27, 2009.

44. For other cases involving humans and animals, see: www.sludgevic-tims.com; sludgefacts.org; www.sludgenews.org; www.sourcewatch.org; and http://sewagesludgeactionnetwork.com/

45. *Ibid, Lewis v. EPA,* Office of Administrative Law Judges, 2003.

46. *Ibid, Lewis et al.,* 2002

47. Gray A, WEF Dep. Exec. Dir., to Whitman C, EPA Admin., copied to Fisher L, Deputy EPA Administrator, Longest H, Acting Asst. Administrator, Mehan T, Cook M, and others, Feb. 13, 2002; Thomas A, Synagro Executive VP and General Counsel, to Fisher L, copied to O'Connor DJ, EPA Acting Asst. Administrator and others, Dec. 10, 2001 [Letter].

48. Weaver GM, Hollberg & Weaver, LLP, to Snell B, US Dept. Justice, Dec. 21, 2012.

49. University of Georgia, Public Affairs Division, Georgia Opens Records Request #14-007, Aug. 8, 2013.

50. FOIA request EPA-HQ-2013-008011 submitted by Attorney James E. Carter on Jul. 7, 2013.

Chapter 4

1. US Dept. Labor, Office of Administrative Law Judges. Case No.99-CAA-12, Deposition transcript of Dr. Alan Rubin, p. 149. *Lewis v. EPA,* Apr. 27, 1999.

2. *Ibid,* pp. 168-172.

3. *USA, ex rel. Lewis, McElmurray and Boyce v. Walker et al.* United States District Court, Middle District of Georgia, Athens Division. Case No. 3:06-CV-16-CDL. Deposition transcript of Dr. Rufus Chaney, Jun. 26, 2009.

4. Dendy LB, U. of Maryland administrator named dean of UGA College of Agricultural and Environmental Sciences. University of Georgia, June 3, 2005.

5. University of Georgia, Bush taps former UGA dean for REE under secretary. Jan. 19, 2006.

6. *Lewis v. EPA,* U.S. Department of Labor, CA 2003-CAA-00005, 00006. Deposition of Robert E. Hodson, Ph.D. Jan. 31, 2003.

7. *Ibid,* Dendy LB, 2005.

8. *Ibid, USA, ex rel. Lewis, et al.* R Chaney Deposition, p. 21.

9. *Ibid, USA, ex rel. Lewis,et al.* R Chaney Deposition, p.53-54.
10. U.S. Environmental Protection Agency. 2002. *Land Application of Biosolids Status Report*; Report 2002-S-000004; Office of Inspector General. Washington, DC.
11. *McElmurray v. United States Department of Agriculture*, United States District Court, Southern District of Georgia. Case No. CV105-159. Order issued Feb. 25, 2008.
12. *Ibid, USA, ex rel. Lewis,et al.* R Chaney Deposition, p. 157.
13. Chaney R, USCC Listserve, October 5, 2004.
14. *Ibid, USA, ex rel. Lewis,et al.* R Chaney Deposition, p. 13-17.
15. Thomas AL, Synagro, to Adams M, President, University of Georgia, Dec. 21, 2004.
16. Synagro Technologies, Inc. Analysis of David Lewis' Theories Regarding Biosolids, p. 4, 6, Sept. 20, 2001.
17. Lewis DL, Gattie DK, Novak ME, Sanchez S, Pumphrey C. Interactions of pathogens and irritant chemicals in land-applied sewage sludges (biosolids). *BMC Public Health* 2:11 (28 Jun). www.biomedcentral.com/1471-2458/2/11.
18. *Ibid*, Lewis *et al.*, *BMC-Public Health*, 2001. Results, Environmental Assessment.
19. Khuder S, Milz SA, Bisesi M, Vincent R, McNulty W, Czajkowski K. Health survey of residents living near farm fields permitted to receive biosolids. *Arch. Environ. Occup. Health* 62, 5–11 (2007).
20. *Ibid*, Lewis *et al.*, *BMC-Public Health*, 2001. Methods, Assessing environmental conditions.
21. Gattie DK and Lewis DL. 2004. A high-level disinfection standard for land-applied sewage sludge (biosolids). *Environ. Health Perspect.* 112:126-31.
22. Reinthaler FF, Posch J, Feierl G, Wüst G, Haas D, Ruckenbauer G, Mascher F, Marth E. Antibiotic resistance of E. coli in sewage and sludge. *Water Res.* 2003 Apr; 37(8):1685-90; Sahlström L, Rehbinder V, Albihn A, Aspan A, Bengtsson B. Vancomycin resistant enterococci (VRE) in Swedish sewage sludge. *Acta Veterinaria Scandinavica*, 2009.
23. *USA, ex rel. Lewis, McElmurray and Boyce v. Walker et al.* United States District Court, Middle District of Georgia, Athens Division. Case No. 3:06-CV-16-CDL. Deposition transcript of J. Gaskin, p. 293-4, Jun. 20, 2009.

24. *Ibid*, p. 269.

25. *Ibid*, p. 372-4, Jun. 22, 2009.

26. Harrison, EZ, McBride MB and Bouldin DR. Land applica-tion of sewage sludges: An appraisal of the US regulations. *Int. J. Environ. and Pollution*, Vol.11, No.1. 1-36. http:cwmi.css.cor-nell.edu/PDFS/LandApp.pdf; Case for Caution Revisited 2009. http:cwmi.css.cornell.edu/case.pdf.http://cwmi.css.cornell.edu/PDFS/LandApp.pdf

27. *Lewis v. EPA*, U.S. Department of Labor, Office of Administrative Law Judges, Case No. 98-CAA-13, Deposition of Ellen Harrison, p. 34-35, 76, Mar. 21, 2003.

28. Harrison E, Cornell Waste Management Institute, to Lewis D, Mar. 5, 2003 [Email].

29. *Ibid*.

30. Kester G to EPA officials Rubin A, Hais A, Roufael A, Carkuff A, Sajjad A, Bastian R, Brobst R, Sans C, Hamilton D, Hetherington D, Gross C, Lindsey A, Home J, Ryan J, Smith J, Colletti J, Dombrowski J, Dunn J, Walker J, Fondahl L, Dominy M, Meckes M, Murphy T. Subject: FW: Dr. David Lewis, 09/24/01 [Email].

31. *Ibid*.

32. National Research Council. *Biosolids Applied to Land: Advancing Standards and Practice*, Overarching Findings, p. 4. National Academy Press. Washington, DC, 2002.

33. Burkhart J (NIH/NIEHS) to Lewis D, copied to Burleigh K (NIH/NIEHS), Subject: EHP ms 6207. May 07, 2003 [Email].

34. *Ibid*, Gattie, Lewis, 2004.

35. Stuck in the mud—The Environmental Protection Agency must gather data on the toxicity of spreading sewage sludge [Editorial]; Tollefson J. Raking through sludge exposes a stink. *Nature*, 2008, Vol. 453, p. 258, 262-3, May 15, 2008.

36. *Nature* editors. *Correction*. *Nature* 453; 577, May 28, 2008 doi:10.1038/453577d   http://www.nature.com/news/2008/080528/full/453577d.html

37. Harrison E. Correspondence at *Nature.com*. June 17, 2008.

38. O'Dette R, Synagro Technologies, Inc., to Stavinoha TD, Commissioner Precinct 1, Fort Bend County, TX. Nov.18, 2002.

39. Personal Communication (2014). McElmurray RA III. Hephzibah, Georgia.

40. Schmitt B, Swickard J. Monica Conyers Gets 37 Months in Prison in Synagro Bribery Scandal in Detroit. *Detroit Free Press - MI*, Mar. 10, 2010; Dixon J. Synagro execs knew of payments, records show. *Detroit Free Press - MI*. Jul. 10, 2010.

41. Lewis DL to Hallman FE Jr. Hallman & Wingate, LLC, Marietta, GA. Nov. 19, 2012 [Letter].

42. Snell B, Civil Division Chief, U.S. Attorney's Office, Middle District of Georgia to FE Hallman, Jr. Hallman & Wingate, LLC. Dec. 12, 2012 [Email].

43. *Ibid.*

44. Olens SS, Attorney General of Georgia, *et al.*, Satisfaction of Judgment and Bills of Costs, *Lewis et al. v. Walker et al.*, U.S. District Court, Middle District Of Georgia, Athens Division, Civil Action File NO. 3:06-cv-16-CDL. Jan. 29, 2013.

45. *Marshall v. Synagro.* State of New Hampshire Superior Court. Case No. 99-C-0045. Deposition transcript of Michael W. Rainey, NH Dept. Environ. Services. Oct. 26, 1999, p. 86-97;114, 123-124; Mehan GT III. USEPA, Assistant Administrator, Office of Water, to Mendelson J III. Dec. 24, 2003 [Letter].

46. For other cases involving humans and animals, see www.sludgevictims.com; sludgefacts.org; www.sludgenews.org; www.sourcewatch. org; and http://sewagesludgeactionnetwork.com/

Chapter 5

1. Gattie DK, Lewis DL. A high-level disinfection standard for land-applied sewage sludge (biosolids). *Environ. Health Perspect.* 2004;112:126-31. Note: PubMed misidentified the second author as McLaughlin TJ. *See*, http://www.ncbi.nlm. nih.gov/ pubmed/14754565. PubMed Canada attempted, unsuccessfully, to have PubMed correct the authorship.

2. Lowman A, McDonald MA, Wing S, Muhammad N, Land Application of Treated Sewage Sludge: Community Health and Environmental Justice. *Environ. Health Perspect.* http://dx.doi. org/10.1289/ehp.1205470. Online Mar. 11, 2013.

3. Lewis DL, Gattie DK, Novak ME, Sanchez S, and Pumphrey C. Interactions of pathogens and irritant chemicals in land-applied sewage sludges (biosolids). *BMC Public Health.* 2002;2:11 (28 June) www.biomedcentral.com/1471-2458/2/11.

4. N Beecher to C Dilworth, National Institute of Environmental Health Sciences (NIEHS), Mar. 2013 [Letter].

5. ABBA Quarterly Conference Call, Mar. 25, 2013. A total of 24 participants included, for example, Lisa McFadden (WEF and the EPA-funded National Biosolids Partnership), Greg Kester (California Association of Sanitation Agencies a.k.a. CASA), Layne Baroldi (CASA, Synagro), Lori Loder (CASA, U.S. Compost Council) and Dan Noble (California Association of Compost Producers).

6. Wing S. When Research Turns to Sludge. *Academe Online*, Nov.-Dec. 2010. http://www.aaup.org/AAUP/pubsres/academe/2010/ND/feat/wing.htm

7. National Biosoilds Partnership. http://www.wef.org/biosolids/

8. N Beecher, NEBRA. Synagro responds to biosolids recycling opponent, *New England Water and Wastewater News*, Nov. 2001; 69:7; N Beecher, NEBRA. David Lewis, Ph.D. Shifts Focus, NEBRA, Apr. 9, 2012; Lewis DL. Institutional Research Misconduct: An Honest Researcher's Worst Nightmare. *Autism Sci. Digest.* April 2012, pp. 33-40.

9. Lewis DL to Carlyle K, National Press Coordinator, UGA Office of Public Information, Nov. 16, 2003 [Email].

10. Deposition transcript of Regina Smith, Ph.D., Apr. 27, 2009, p.79. *USA ex rel., Lewis et al. v. Walker et al.* U.S. District Court, Middle District of Georgia, Athens Division. Case No. 3:06-CV-16.

11. Holmes C. University of Georgia. Sludge study relieves environmental fears, Jan. 29, 2003.

12. Environmental Public Health Review prepared by Louisiana Department of Health and Hospitals, Jul. 19, 2005; Harrison E to The Rubins...re: Louisiana Convent—more sludge victims? Aug. 12, 2005 [Email].

13. Farfel MR, Orlova AO, Chaney RL, Lees PSJ, Rohde C and Ashley PJ. Biosolids compost amendment for reducing soil lead hazards: AA pilot study of Orgro® amendment and grass seeding in urban yards. *Sci. Total Environ.* 2005;340: 81-95.

14. Southwestern Illinois Resource Conservation & Development, Inc., East St. Louis Biosolids Lead Remediation Project funded by US EPA, USDA Natural Resources Conservation Service. http://www.swircd.org/pdf/lead.pdf.

15. *Ibid*, Gattie DK, Lewis DL, 2004.

16. *Lewis v. EPA*. U.S. Department of Labor, Office of Administrative Law Judges, Case Nos. 2003-CAA-00005, 2003-CAA-00006. Transcript of Hearing before the U.S. Department of Labor Northeast Region in Washington, DC, Apr. 11, 2003.

17. *Ibid*.

18. Heilprin J, Vineys KS. *AP IMPACT*: Sludge tested in poor neighborhoods, Associated Press, Baltimore, MD, Apr. 13, 2008.

19. *USA ex rel., Lewis et al. v. Walker et al.* Deposition transcript of Rufus Chaney, Ph.D. Jun. 26, 2009, p. 69;

20. Personal Communication. Heilprin J, Mar. 29, 2009.

21. Chaney R, USDA, to Gaskin J, Angle JS and others. Nov. 20, 2008 [Email].

22. Silverman M, Managing Editor, AP, to Hallman FE Jr., copied to Heilprin J, Tomlin D. Subject: Request for information regarding AP reporter John Heilprin, Feb. 11, 2009 [Email].

23. *Ibid*, Deposition transcript of R Chaney, 2009.

24. *Ibid*.

25. United Nations Correspondent Association, Announcing the Thirteenth Annual UNCA Award Winners for 2008, Elizabeth Neuffer Memorial Award for Best Overall Print Journalism, including online media, Dec. 4, 2008. http://cms.unca.com/content/ view/42/10/.

26. Lewis DL to Snowden CO, Stansbury GG, Cheatham ML Sr, Johnson ME. Subject: Lewis, McElmurray, Boyce cases, Johns Hopkins/Kennedy Krieger experiments. Jun. 8, 2008. 9 pp [Letter].

27. Cecil KM, Brubaker CJ, Adler CM, Dietrich KN, Altaye M., *et al.* Decreased Brain Volume in Adults with Childhood Lead Exposure. *PLoS Med.* 5: e112 (2008); Wright JP, Dietrich KN, Ris MD, Hornung, RW, Wessel SD, *et al.* Association of Prenatal and Childhood Blood Lead Concentrations with Criminal Arrests in Early Adulthood. *PLoS Med.* 5: e101 (2008).

28. *Federal Register*, 2001. Part III, U.S. Environmental Protection Agency. Lead; Identification of dangerous levels of lead: Final Rule 66, 1206-1240.

29. Snyder C. Baltimore Sludge Pilot Project Puts Children at Additional Risk. 2008; *Int J Occup Environ Health*. 14 (3) 240-241. http://www.sludgefacts.org/baltimore.pdf

30. Binns HJ, Gray KA, Chen T, Finster ME, Peneff N, Schaefer P, Ovsey V, Fernandes J, Brown M, and Dunlap B. Evaluation of

landscape coverings to reduce soil lead hazards in urban residential yards: The Safer Yards Project. *Environ Res.* 2004;96:127-38.

31. U.S. Environmental Protection Agency, Office of Pollution Prevention and Toxics. 2000. Analysis of pathways of residential lead exposure in children. EPA 747-R-98-007.

32. Ryan, JA, Berti WR, Brown SL, Casteel SW, Chaney RL, Doolan M *et al.*, Reducing children's risk from soil lead. Summary of a field experiment. *Environ. Sci. Technol.* 2004;38:18A-24A. http://www.esd.ornl.gov/research/earth_sciences/images/2004ryan_esd_38_10A-24A.pdf

33. Heilprin J, Vineys KS, Senate Plans Hearing on Sludge. *Associated Press*, April 14, 2008.

34. Senator Barbara Boxer to David L. Lewis, Sept. 4, 2008 [Letter].

35. Hallum AW. Compliance Evaluation Task Force, Georgia Department of Natural Resources, Georgia Environmental Protection Division, to Larson JH, Sommerville, GA, Dec 16, 1998 [Memorandum].

36. U.S.EPA, Lewis DL to Holm H, Athens-ERL Research Director. Adverse health effects from Augusta-sludged hay. May 8, 2003 [Memorandum].

37. Lee J, Sludge Spread on Fields Is Fodder for Lawsuits, *NY Times*, Jun. 26, 2003.

38. *United States of America, ex rel. David L. Lewis, Ph.D., R. A. McElmurray, III, and G. William Boyce v. John Walker, Ph.D., Julia W. Gaskin, Robert B. Brobst, William P. Miller, Ph.D., E. William Tollner, Ph.D., L. Mark Risse, Ph.D., Joe. L. Key and The University of Georgia Research Foundation, Inc.* United States District Court, Middle District of Georgia, Athens Division. Case No. 3:06-CV-16-CDL.

39. *USA ex rel., Lewis et al. v. Walker et al.* Deposition transcript of William P. Miller, Ph.D., Jan. 22, 2009, p.134.

40. *Ibid*, Holmes 2003.

41. Lewis DL to Leed AH, UGA Associate Director of Legal Affairs, Aug. 10, 2005 [Memorandum].

42. Leed AH to Kohn SM, Jan. 19, 2006 [Letter].

43. Smith R, UGA Scientific Integrity Officer, to Mace A, UGA Provost, and Patel G, UGA VP Research, Apr. 19, 2004 [Memorandum].

44. Gaskin J to Hallman FE Jr., Apr. 21, 2004 [Letter].

45. Evans B, Senate cancels hearing on Georgia sludge findings. *Associated Press*, Sept. 11, 2008.

46. *USA ex rel., Lewis et al. v. Walker et al.*, DL Lewis ("Plaintiff ") Responses to Dr. Joe L. Key and University of Georgia Research Foundation, Inc., Requests for Production, Requests for Admission, May 4, 2009.
47. *USA ex rel., Lewis et al. v. Walker et al.*, Deposition transcript of David Lewis, Ph.D., Vol III, Jun. 22, 2009, p. 349.

Chapter 6

1. Bingham J, "Fracking protesters like MMR scaremongers, says Church of England," *The Telegraph*, Aug. 16, 2013.
2. HHS.gov.Nov.14,2011. Faith-based and Neighborhood Partnerships, Tools and Resources, *The Partnership Center Newsletter*.
3. *Ibid.*
4. National Institute of Environmental Health Sciences, Environmental Health Disparities and Environmental Justice Meeting, Research Triangle Park, NC, Jul. 29-31, 2013.
5. Kuehn R. (2004). Suppression of Environmental Science, Section III, Part B. Misuse Of Scientific Misconduct Charges (pp. 349-55). *American Journal of Law & Medicine* 30:333-69.
6. Wheatly MJ, *Turning to One Another*. Berrett-Koehler Publishers, Inc., 2002.
7. Sekerka L. Appreciative Inquiry Organizational Development Program, Naval Postgraduate School, Monterey, CA, Jan. 10, 2003.
8. Kamen A. "Intimate Climate." *Washington Post*, p. A19, Jan. 10, 2003.
9. Petersen M, Beef Cattle become Behemoths: Who Are Animal Scientists Serving? Apr. 15, 2013. *Chronicle of Higher Education.* http://chronicle.com/article/As-Beef-Cattle-Become/131480/
10. Michaels D, Doubt is their product: How industry's assault on science threatens your health. 2008 Introduction. Oxford University Press, New York.
11. *James* 1:27.
12. *Quran* 3.119; 2.190-91.
13. *Hebrews* 4:12.
14. *Deuteronomy* 27:26; *James* 2:10; *Quran* 3.119.
15. *James* 2:10.
16. *Romans* 1:21 to 2:1-6.
17. *Exodus* 16:3.
18. *Exodus* 34:6-7.

19. *James* 2:13.
20. *John* 8:7.
21. *Matthew* 7:2.
22. *Matthew* 7:12.
23. *James* 2:19.
24. *Matthew* 7:21.
25. *Isaiah* 14:14.
26. *Isaiah* 19:25.
27. *Mark* 9:24; *John* 20:25.
28. *Matthew* 7:12; 22:37-40.
29. *1 John* 4:7.
30. *Matthew* 25:31-46
31. *Jeremiah* 31:33; *Hebrews* 10:16
32. *John* 3:7.
33. *Deuteronomy* 1:39.
34. *Genesis* 2:9-25.
35. *James* 1:27.
36. *Matthew* 25:31-46.

Chapter 7

1. The White House, "We the People petition to work with the new EPA administrator to ban the land application of sewage sludge (also called biosolids)." Snyder C, Dec. 27, 2012.
2. Wakefield AJ, Murch SH, Anthony A, Linnell J, Casson DM, Malik M, Berelowitz M, Dhillon AP, Thomson MA, Harvey P, Valentine A, Davies SE, Walker-Smith JA. Ileal lymphoid nodular hyperplasia, non-specific colitis and pervasive developmental disorder in children, *Lancet* 1998;351(9713):637-641. [Retracted]
3. Hsiao EY, McBride SW, Hsien S, Sharon S, Hyde ER, McCue T, Codelli JA, Chow J, Reisman SE, Petrosino JF, Patterson PH, Mazmanian SK. Microbiota Modulate Behavioral and Physiological Abnormalities Associated with Neurodevelopmental Disorders. *Cell*, 05 December 2013; doi:10.1016/j.cell.2013.11.024; Walker SJ, Fortunato J, Gonzalez LG, Krigsman A (2013). Identification of Unique Gene Expression Profile in Children with Regressive Autism Spectrum Disorder (ASD) and Ileocolitis. *PLoS ONE* 8(3): e58058. doi:10.1371/journal.pone.0058058; Williams BL, Hornig M, Buie T, Bauman ML, Cho Paik M, Wick I, Bennett A, Jabado O,

Hirschberg DL, Lipkin WI. Impaired carbohydrate digestion and transport and mucosal dysbiosis in the intestines of children with autism and gastrointestinal disturbances. *PLoS One.* 2011 Sep;6(9): e24585.

4. Deer B, Revealed: MMR research scandal, *The Sunday Times*, February 22, 2004.

5. General Medical Council, Committee and Professional Conduct Committee (UK). Dr. Andrew Jeremy Wakefield and Dr. John Walker-Smith, Determinations on Serious Professional Misconduct (SPM) and Sanction, May 24, 2010.

6. *Professor John Walker-Smith v. General Medical Council, Committee and Professional Conduct Committee.* The High Court of Justice, Queen's Bench Division, Administrative Court. Case No: CO/7039/2010. March 3, 2012.

7. *Dr. Andrew J. Wakefield v. The British Medical Journal, Brian Deer and Fiona Godlee,* District Court for the 250th District, Travis, TX, Case No. D-1-GN-12-000003, January 3, 2012.

8. Synagro Technologies, Inc. Analysis of David Lewis' Theories Regarding Biosolids, including separate Executive Summary. Sept. 20, 2001.

9. *Lewis v. EPA,* U.S. Department of Labor, Case Nos. 2003-CAA-6, 2003-CAA-5, Joint stipulations, March 4, 2003; USEPA, Stancil F, Branch Chief, ESD, to Russo R, Director, ERD, Apr. 22, 2003; Thomas AL II, Synagro Exec. VP & Gen. Counsel, to Adams MF, President, University of Georgia. Dec. 21, 2004 [Letter].

10. Lewis DL, Letter to the *BMJ* from David Lewis, Rapid response, http://www.bmj.com/rapid-response/2011/11/09/re-how-case-against-mmr-vaccine-was-fixed; Godlee F, "Institutional research misconduct," *BMJ* 2011; Nov 9;343:d7284, http://www. bmj.com/content/343/bmj.d7284?tab=full. Additional documents are posted under research staff (David Lewis) by the University of British Columbia Neural Dynamics Research Group. http:// www.neuraldynamicsubc.ca/ .

11. Deer B, "David L Lewis: Indignant abuse as complaints turn to nothing," Jan. 10, 2012. http://briandeer.com/solved/david-lewis-1.htm.

12. Deer B, Scientific misconduct: Latest MMR 'dispute' is a straw man. *Nature* 481, 145, Jan.12, 2012.

13. Godlee F. *BMJ* response to David Lewis' accusations of scientific fraud, Jan. 23, 2012.

14. Hallman FE Jr. to Godlee F, Feb. 4, 2013 [Letter].

15. Deer B to Carter JE Apr 22, 2013 [Email].

16. *Ibid.*

17. Vaccine Safety: Evaluating the Science. Montego Bay, Jamaica Jan. 3-7, 2011.

18. CNN Anderson Cooper 360° Jan 6, 2011; CNN's Dr. Sanjay Gupta questions Andrew Wakefield; CNN Danielle Dellorto, February 4, 2011, Bill Gates: Vaccine-autism link 'an absolute lie'; CNN, Journalist Brian Deer responds to Dr. Andrew Wakefield, Jan. 7, 2011.

19. *Ibid*, Wakefield *et al.*, *Lancet*, 1998.

20. Deer B, Wakefield's "Autistic Enterocolitis" under the microscope. *BMJ* 2010;340:c1127.

21. *Ibid.*

22. Mikhail N, Lewis DL, Omar N, Taha H, El-Badawy A, Mawgoud NA, Abdel-Hamid M, Strickland GT. Prospective study of cross-infection from upper-GI endoscopy in a hepatitis C–prevalent population. *Gastrointest Endosc* 2007;65:584-588.

23. General Medical Council (GMC) Fitness to Practise Panel (Misconduct). Transcript of the shorthand notes of TA Reed & Co., Ltd. (2010), Professor John Angus Walker-Smith Examined By Mr. Miller, Day 74-38, July 16, 2008.

24. Dhillon AP. Re: Pathology reports solve "new bowel disease" riddle. *BMJ*, Nov. 17, 2011.

25. *Ibid*, GMC Walker-Smith testimony, Day 74-37.

26. *Ibid*, AP Dhillon. *BMJ*, 2011.

27. *Ibid*, Godlee, *BMJ* Nov. 9, 2011.

28. Godlee F, Smith J, Marcovitch H. Wakefield's article linking MMR vaccine and autism was fraudulent. *BMJ* 2011; 342:c7452.

29. Booth I, Expert Witness Report submitted to General Medical Council, Fitness To Practice Panel (Misconduct), Nov. 8, 2006.

30. Deer B to Kohn S, Jun. 2, 2011 [Email].

31. *Ibid*, Booth I 2006.

32. Booth to Lewis DL, Aug. 10, 2011 [Email].

33. *Ibid*, Godlee et al., *BMJ* 2011.

34. Reich ES. "Fresh dispute about MMR 'fraud,'" *Nature* 2011;479:157-158.

35. Deer B. "Brian Deer Wins a Second British Press Award" and "Award-Winning Journalism at *The Sunday Times*." http://briandeer. com/brian/press-awards-2011-win.htm

36. Deer B. Wakefield's "Autistic Enterocolitis" under the microscope. *BMJ* 2010;340:c1127.

37. *Ibid*, Deer to Kohn, 2011.

38. *Ibid*, Reich 2011.

39. *Ibid*.

40. *Ibid*, Booth 2011.

41. *Ibid*, Deer 2010.

42. Delamothe T to Lewis DL, copied to Godlee F. Oct. 27, 2011 [Email].

43. Lewis DL, "Letter to the *BMJ* from David Lewis," *BMJ* Rapid response, 2011.

44. *Professor John Walker-Smith v. General Medical Council, Committee and Professional Conduct Committee*. The High Court of Justice, Queen's Bench Division, Administrative Court. Case No: CO/7039/2010. Mar. 3, 2012.

45. University College London, "MMR and the development of a research governance framework in UCL," Sep. 13, 2012. http://www. ucl.ac.uk/news/news-articles/1209/13092012-Governance.

46. Thompson N, Montgomery S, Pounder RE, Wakefield AJ. Is measles vaccination a risk factor for inflammatory bowel diseases? *Lancet* 1995; 345: 1071–74.

47. Davis RL, Bohlke K. Measles vaccination and inflammatory bowel disease: controversy laid to rest? *Drug Saf.* 2001;24(13):939-46.

48. Williams BL, Hornig M, Buie T, Bauman ML, Paik M Cho, Wick I, Bennett A, Jabado O, Hirschberg DL, Lipkin WI. Impaired carbohydrate digestion and transport and mucosal dysbiosis in the intestines of children with autism and gastrointestinal disturbances. *PLoS One*. 2011 September;6(9): e24585.

49. Ibid, Walker *et al.*, 2013.

50. *Ibid*, CNN Anderson Cooper 360°, 2011.

Chapter 8

1. Booth I to Lewis DL, Subject: Booth expert report to GMC, Fitness to Practice Panel (Misconduct), "Wakefield, Walker-Smith, Murch. Second Addendum to Overview Statement," Aug. 10, 2011 [Email].

2. Deer B, Wakefield's "Autistic Enterocolitis" under the microscope. *BMJ* 2010;340:c1127.

3. Lohn M, Field Fisher Waterhouse Partner, to Deer B, Subject: General Medical Council - Dr Wakefield, Dr Murch, Dr Walker-Smith, Reference No. JXO/00492-I 5237/3352872 v1, May 25, 2005.

4. Godlee F, Smith J, Marcovitch H. Wakefield's article linking MMR vaccine and autism was fraudulent. *BMJ* 2011; 342:c7452.

5. *Ibid.*

6. Godlee F. "Institutional research misconduct." Editorial published in response to Lewis DL *Rapid Response* letter. *BMJ* 2011; Nov 9;343:d7284. Disclaimer states: Author declares "no financial relationships with any organisations that might have an interest in the submitted work in the previous three years; the BMJ Group receives funding from the two manufacturers of MMR vaccine, Merck and GSK."

7. Godlee F. Uncorrected Transcript of Oral Evidence, House of Commons Science and Technology Committee, Peer Review, May 11, 2011. http://www.publications.parliament.uk/pa/cm201012/cmselect/cmsctech/uc856-ii/uc85601.htm.

8. Deer B. "Brian Deer Wins a Second British Press Award" and "Award-Winning Journalism at *The Sunday Times*." http://briandeer.com/brian/press-awards-2011-win.htm

9. *Ibid*, Godlee *et al.* 2011. Lewis DL. "Wakefield Fights Back" submitted to *Nature Neuroscience* on Mar. 2, 2011. Author's manuscript cited an Alliance for Human Research Protection Partnership article, "*BMJ* & *Lancet* Wedded to Merck CME" (http://www.ahrp.org/cms/content/view/766/149/). It also included a figure illustrating GSK's sponsorship of *BMJ* (http://groupawards.bmj.com/sponsors), including a figure legend stating: "Financial ties between medical journals and MMR vaccine manufacturers. Are they playing a role in attacks against Wakefield?" In a "report" to the BMJ published in Jan. 2012 (http://briandeer.com/solved/david-lewis-1.htm), Brian Deer described first learning about me when he was informed about a manuscript I submitted for publication, which was rejected by a U.S. journal. In the manuscript, which I submitted to an ethics journal, I discussed the BMJ's and, in turn, Deer's conflicts of interest with manufacturers of MMR vaccines. John Stone of Age of Autism

- UK had raised similar concerns directly to the BMJ in March 2011 (http://www.bmj.com/content/342/bmj.d1335?tab=responses).

10. *Andrew Wakefield v. Channel Four Television Corporation, Twenty Twenty Productions, Ltd. and Brian Deer,* High Court of Justice, Queen's Bench Division, Case No. HQ05X00900.

11. *Professor John Walker-Smith v. General Medical Council, Committee and Professional Conduct Committee.* The High Court of Justice, Queen's Bench Division, Administrative Court. Case No: CO/7039/2010. Mar. 3, 2012.

12. *Ibid.* at 15, 23.

13. *Ibid.* at 20.

14. *Ibid.* at 24-49.

15. *Ibid.* at 50.

16. *Ibid.* at 85, 186.

17. *Ibid.* at 148, 186.

18. *Ibid.* at 16.

19. *Ibid.* at 153.

20. Kraus S, London Strategic Health Authority, National Health Services (NHS) to Miller CG, (Jan. 15, 2007). A description by DL Lewis of the 10 documents NHS provided to Deer follows: (1) [Letter] John Walker-Smith to Maureen Carroll, Royal Free Hospital Ethics Committee (1997 February 27) in which J Walker-Smith states: "We currently have formal approval to take research biopsies during colonoscopy (code 162-95) and I am writing to organize formal approval for research biopsies to be taken during upper biopsies," (2) [Letter] AD Phillips to M Pegg, Royal Free Hospital and Medical School Ethics Committee (2000 March 15) in which Phillips requests an "updated approval" for continuing to take research biopsies in studies 162-95 (colonoscopy) and 70-97 (upper endoscopy), (3) Same as No. 2 above with handwritten note from Pegg to Carroll requesting she check on the status of the approvals, (4) [Memorandum] Carroll to Pegg (2000 April 7) re. "Extension to 162-95 & 70-97," in which Carroll states: "...no requests have been received previously for the continuation of these studies...could you please approve this by way of Chairman's Action and return to me," (5) [Letter] Pegg to Phillip (2000 April 28) in which Pegg acknowledges Phillip's March 15 letter and requests a brief annual report for studies 162-95 and 70-97, (6) [Letter] Phillips to Pegg (2000 May 17), which is a cover letter transmitting

"1999 Annual Report on Ethical Submissions 162-95 and 70-97," (7) [Report] Phillips to Pegg (2000 May 17), "1999 Annual Report on Ethical Submissions 162-95 and 70-97," (8) Same as No. 6 above with handwritten note from Pegg to Carroll requesting she "create an updated approval letter" and stating: "I think I should sign it," (9) [Letter] Pegg to Phillips (2000 May 25), in which Pegg acknowledges receiving the annual report on the taking of research biopsies, and conveys the committee's approval for Walker-Smith's group to continue taking research biopsies in studies 162-95 and 70-97, and (10) a blank, undated parental consent form titled "CONSENT FORM FOR RESEARCH BIOPSIES."

21. TA Reed & Company, Day 68-57 thru 59.
22. *Ibid*, TA Reed & Company, Day 197-2 and 3; General Medical Council, Committee and Professional Conduct Committee (UK) (2010 May 24), "Dr. Andrew Jeremy Wakefield—determination on serious professional misconduct (SPM) and sanction," http://www. gmc-uk.org/   Wakefield_SPM_and_SANCTION.pdf_32595267. pdf; and General Medical Council, Committee and Professional Conduct Committee (UK). (2010 May 24), "Professor John Angus Walker-Smith—determination on serious professional misconduct (SPM) and sanction," http://www. gmc-uk.org/Professor_Walker_Smith_SPM.pdf_32595970.pdf.
23. *Ibid*, TA Reed & Company. Day 147-62.
24. AD Phillips to MS Pegg, Royal Free Hospital Medical School Ethics Committee (2000 May 17).
25. *Ibid*, *Walker–Smith v. GMC* at 7.
26. JS Brown, BP Kotler, RJ Smith, WO Wirtz II, The effects of owl predation on the foraging behavior of heteromyid rodents, *Oecologia* 1988; 76:408-415.
27. Personal communication; Wakefield A to Horton R, Mar. 15, 2012 [Letter].
28. *Ibid*, Godlee F. Uncorrected Transcript of Oral Evidence, 2011.

Chapter 9

1. "Research Ethics: A Question of Method?" Harvard University John F. Kennedy School of Government. Cosponsored by the Program on Science, Technology and Society and Cultural Foundations of

Integration, a Center of Excellence at the University of Konstanz, Baden-Württemberg, Germany. May 6, 2011.

2. *Lewis v. EPA*, U.S. Department of Labor (DOL). In Case No. 97-CAA-7, 1997, for example, the DOL found: "Dr. David Lewis was discriminated against by EPA's inquiry into ethics violations at the highest levels and communicating these allegations to members of Congress." EPA paid $115,000 in damages and issued a letter stating that I did not violate the Hatch Act or any ethics rules. In Case No. 41107317, 1998; and Case Nos. 1999-CAA12; 2000-CAA10 and 11, the DOL ruled that EPA retaliated over my *Nature* commentary, for example, by establishing special requirements for my promotion to GS 15. EPA paid $495,000 in damages and legal costs to settle.

3. NIH Committee on Scientific Conduct and Ethics, "A Guide to the Handling of Scientific Misconduct Allegations in the Intramural Research Program at the NIH," January 12, 2001. http://sourcebook.od.nih.gov/ResEthicsCases/sm-booklet.htm

4. Godlee F. *BMJ* response to David Lewis' accusations of scientific fraud, Jan. 23, 2012. http://www.bmj.com/rapid-response/2012/01/23/re-how-case-against-mmr-vaccine-was-fixed; B Deer, "David L Lewis: indignant abuse as complaints turn to nothing," Jan. 10, 2012. http://briandeer.com/solved/david-lewis-1.htm.

5. Walker JM, EPA Municipal Technology Branch, to Longest HL II, EPA Assoc. Deputy Asst. Administrator for Water Program Operations. September 12, 1978.

6. "Stuck in the Mud." *Nature* 453: 258; 262-3, May 15, 2008 [Editorial].

7. U.S. EPA, Technical Assistance Directory, Office of Research and Development. EPA/600/K-97/001 October 1997.

8. Paletta D, White House to Halt Civil-Service Bonus Program, *WSJ*, Jun. 11, 2013.

9. Personal communication.

10. House Bill 3288 Effective 09-01-99, which amended the Solid Waste Disposal Act to "prohibit the Texas Natural Resource Conservation Commission (TNRCC) from charging a solid waste disposal fee for the disposal of sewage sludge that has been treated to reduce the density of pathogens to the lowest level provided by TNRCC rules...."

11. *Marshall v. Synagro*, Rockingham County, New Hampshire Superior Court, Case No. 99-C-45. Deposition of David L. Lewis, August 9, 2001, p. 800-801.

12. *Lewis v. EPA*, U.S. Dept. Labor Case Nos. 2003-CAA-00005, 00006. Deposition of Rosemarie C. Russo, Ph.D., Jan. 31, 2003; Woods J, EPA ORD, to Lewis D, "Emergency Access to Scientific Advice." Jun. 20, 2002 [Email]; Lewis DL to Russo R, EPA Homeland Security, Jun. 3, 2002 [Memorandum].

13. *USA, ex rel. David L. Lewis, Ph.D., R. A. McElmurray, III, and G. William Boyce v. John Walker, Ph.D., Julia W. Gaskin, Robert B. Brobst, William P. Miller, Ph.D., E. William Tollner, Ph.D., L. Mark Risse, Ph.D., Joe. L. Key and The University of Georgia Research Foundation, Inc.* U.S. District Court, Middle District of Georgia, Athens Division. Case No. 3:06-CV-16-CDL. Affidavit of David L. Lewis, Ph.D., Exhibit A, Oct. 28, 2009, p. 85.

14. Reich ES, "Agencies unveil plans to safeguard science." *Nature* 476:262, August 18, 2011.

15. *Ibid.*

16. The term "banished from Woolworths" is quoted from the 2000 comedy film, *O Brother, Where Art Thou?*, starring George Clooney.

17. Mann A. Fight over Sludge Starts to Get Dirty. *Time*, Sept. 27, 1999; Follow-Up Mann A. More Sludge Slinging: How Safe if That Dump? *Time*, Oct. 4, 1999.

18. NIOSH. 2002. Guidance for Controlling Potential Risks to Workers Exposed to Class B Biosolids. NIOSH Publication No. 2002-149. Cincinnati, OH: National Institute for Occupational Safety and Health. www.cdc.gov/niosh/docs/2002-149/pdfs/2002- 149.pdf.

19. U.S. House of Representatives, Committee on Science. 2000a. EPA's Sludge Rule: Closed Minds or Open Debate? No. 106-95. Washington, DC: U.S. Government Printing Office; 2000b. Intolerance at EPA— Harming People, Harming Science? No. 106- 103. Washington, DC: U.S. Government Printing Office.

20. Notification and Federal Employee Antidiscrimination and Retaliation Act of 2002. 2002. Public Law 107-174.

21. Shaw GJ. SEA secures vital changes to "no fear" bill. *Senior Exec.* Legislative Update. May 2002, p 3.

22. EPA-WEF *National Biosolids Public Acceptance Campaign*, Grant No. CX-820725-01-0, Decision Memorandum from Michael J. Quigley to Michael B. Cook, July 28, 1992.

23. *Ibid.*

24. Epstein E to Ozonoff D, Chair, Dept. Environ. Health, Boston University School of Public Health. Sept. 28, 2001.

25. USEPA Office of Inspector General, Report of Audit. Management of Extramural Resources. Audit Report E1JBF2-04-0300-3100156 (revised) March 31, 1993; U.S. Government Printing Office. EPA contracting: hearing before the Subcommittee on Oversight and Investigations of the Committee on Energy and Commerce, House of Representatives, One Hundred Third Congress, first session, March 10, 1993, Volume 4; *EPA Watch*: "Browner admits EPA's accountability in disrepair, pledges reform." Vol 2, No 6. March 31, 1993. Available at http://www.legacy. library.ucsf.edu/documentStore/i/a/p/iap56e00/Siap56e00.pdf

26. *USA, ex rel. Lewis, McElmurray and Boyce v. Walker et al.* Deposition of Joe. L. Key, Apr 4, 2009, p. 34-35.

27. Based on Author's records obtained under the Georgia Open Records Act, August 2013.

28. National Center for Biotechnology Information, Colleges of agriculture at the land grant universities: Public service and public policy. *Proc. Natl. Acad. Sci.* Vol. 94, pp. 1610–1611, Mar. 1997.

29. *Lewis v. EPA*, U.S. Department of Labor. CA 2003-CAA-00005, 00006. Deposition Transcript of Robert E. Hodson, Ph.D., Jan. 31, 2003.

Chapter 10

1. Atlanta *Journal-Constitution*, "CDC offers buyouts in chief's office," July 24, 2008.

2. Lewis DL. 1996. EPA Science: Casualty of election politics. *Nature* 381:731-2.

3. Union of Concerned Scientists, Cambridge, MA. Hundreds of EPA scientists report political interference over last five years. Apr. 23, 2008.

4. *Ibid.*

5. National Institute of Environmental Health Sciences, Environmental Health Disparities and environmental Justice Meeting, Research Triangle Park, NC, Jul. 29-31, 2013.

6. U.S. House of Representatives, Committee on Science. 2000a. EPA's Sludge Rule: Closed Minds or Open Debate? No.106-95. Washington, DC:U.S. Government Printing Office.—. 2000b. Intolerance at EPA—Harming People, Harming Science? No. 106-103. Washington, DC:U.S. Government Printing Office.

7. National Research Council. 2002. Biosolids Applied to Land: Advancing Standards and Practice. Washington, DC: National Academy Press.

8. Rockefeller A, Civilization and sludge: notes on the history of the management of human excreta. Current World Leaders, vol.39. No.6. 1996.

9. Snyder C. 2005. The Dirty Work of Promoting Recycling of America's Sewage Sludge. IJOEH 2 (4): 415-27. www.sludgefacts.org

10. Personal communication.

11. Lewis DL. 1999. High-level disinfection of flexible endoscopes: a microbiologist's point of view. International Review of Modern Surgery. pp. 77-83. Published in association with the International Federation of Surgical Colleges.

12. Jensen MN. Bacteria may hide in hunks of gunk. Feb. 28, 1998; Science News, 153:137.

13. Gingrich N, Speaker, U.S. House of Representatives, Norwood C, Member of Congress, Linder J, Member of Congress, to Huggett R, EPA Asst. Administrator ORD, Oct 30, 1996; Senators Inhofe J, Grassley C to Whitman CT, EPA Administrator, May 21, 2003 [Letter].

14. Lee G, Agency takes a hit from one of its own. *Washington* Post, Jun. 27, 1996; Mann A. Fight over Sludge Starts to Get Dirty. *Time*, Sept. 27, 1999; Follow-Up—Mann A. More Sludge Slinging: How Safe if That Dump? Time, Oct. 4, 1999; Faced with faulty science, EPA muzzles critics, USA Today, Oct 5, 2000 [Editorial]; Barnett M, Making a Stink. *U.S. News & World Report*, Aug. 5, 2002.

15. American Dental Association. ADA Urges HHS to Act on Research. *ADA News.* Mar. 9, 1992; Breo D. The dental AIDS cases – murder or an unsolvable mystery? *JAMA* 1993, 270:2732-34; Finch S. Unclean instruments. *Hippocrates.* February 1998; Davis R, Medicine's dirty little secret. USA Today, Feb. 18, 1999; Underwood A, Do scopes

spread sickness? *Newsweek*, Mar. 1, 1999; Armbrister T, Weird Science at the EPA, *Reader's Digest*, June 1999; Appleby J, Medical community debates scope-cleaning procedures, *USA Today*, Apr. 8, 2002; Tollefson J. Raking through sludge exposes a stink, *Nature* 2008;453:262-3; Stuck in the mud—The Environmental Protection Agency must gather data on the toxicity of spreading sewage sludge [Editorial], *Nature* 2008;453:258; Reich ES, Fresh dispute about MMR 'fraud,' *Nature* 2011; 479: 157-158.

16. Lewis DL. 1999. A sterilization standard for endoscopes and other difficult to clean medical devices. *Practical Gastroenterology* 23:28-56.
17. Tucker RC, BS Lestini, RE Marchant. Surface analysis of clinically used PFTE endoscopic tubing treated by the STERIS PROCESS. 1996; ASAIO *Journal* 42:306-313; CDC, Guideline for Disinfection and Sterilization in Healthcare Facilities, 2008.
18. Mikhail N, DL Lewis, N Omar, *et al.*, 2007. 5-year cohort study of cross-infection from upper-GI endoscopy in a hepatitis C–prevalent population. *Gastrointest. Endoscopy.* 65: 584-588; Lewis DL, and M Arens. 1995. Resistance of microorganisms to disinfection in dental and medical devices. Nature Med. 1:956-958; Lewis DL, M Arens, R Harlee, G Michaels. Risks of Infection with Blood- and Saliva-Borne Pathogens from Internally Contaminated Impressions and Models. *Trends & Techniques*, National Association of Dental Laboratories, Jun. 1995.
19. *Ibid*, Mikhail *et al*, 2007.

Chapter 11

1. *Erika Crimes v. Kennedy Krieger Institute, Inc., Circuit Court for Baltimore City*, Case Nos. 24-C-99-000925, 24-C-95066067/CL193461. Order dated Oct. 11, 2001. http://www.courts.state.md.us/opinions/coa/2001/128a00.pdf
2. *Ibid.*
3. *Ibid.*
4. Farfel MR, Orlova AO, Chaney RL, Lees PS, Rohde C, Ashley P. 2005. Biosolids compost amendment for reducing soil lead hazards: A pilot study in urban yards. *Science of the Total Environment* 340:81-95.
5. *Ibid, Crimes v. Kennedy Krieger*, 2001.
6. Snyder C, 2014. Dr/ Snyder documented her recollections specifically for this book, and are not published elsewhere.

7. Gerlach W. And the Witnesses Were Silent: The Confessing Church and the Persecution of the Jews. Translated and Edited by VJ Barnett. University of Nebraska Press. Lincoln and London. 2000.

8. de Zayas AM. *Nemesis at Potsdam: The Expulsion of the Germans from the East.* Third Edition. 1988. University of Nebraska Press. Lincoln and London.

9. Hunt L. Secret Agenda: The United States Government, Nazi Scientists, and Project Paperclip 1945-1990. 1991. St. Martin's Press. New York.

10. Transcript of the Nuremberg Trials. Harvard University. http://nuremberg.law.harvard.edu/NurTranscript/TranscriptPages/411_372.html

11. Martin DS. Vets Feel Abandoned After Secret Drug Experiments. *CNN Health.* Mar. 1, 2012. http://www.cnn.com/2012/03/01/health/human-test-subjects/

12. Joint House and Senate Government Oversight Hearings titled "Activities of the EPA's Office of Inspector General [OIG]" were held in 1991. Rep. John Dingell and Senator John Glenn pressured Martin to step up prosecutions of EPA managers and contractors over contract abuses. Martin, in turn, rewrote performance standards for his field agents, requiring that they achieve certain levels of prosecutions to gain satisfactory performance ratings and promotions. By networking with EPA employees, the Author built a large file documenting the OIG's abuses, which he provided to *Reader's Digest* editor Trevor Armbrister, and Jeff Nesmith in the Washington, DC Bureau of the *Atlanta Constitution*. Martin resigned several weeks after the Author informed EPA of the pending news coverage. Martin's abuses were outrageous. For example, a chemist working at the EPA lab in Athens, Georgia, Dr. Jesse MacArthur (Mac) Long, and his wife, were awakened late one night as agents sent by EPA's Office of Inspector General and the U.S. Department of Justice knocked on their front door. After informing Dr. Long that he was about to be prosecuted, they offered to let him off the hook if he would wear a hidden microphone to gather information from his laboratory director. Martin promised his agents extra points for prosecuting high-profile cases that captured congressional and media attention. According to documents the Author obtained from Dr. Long, the Georgia Bureau of Investigation, Mac planned question his EPA lab director as she took a smoking break outside the building. To protect Dr. Long, GBI agents promised to be hiding in the shrubbery and

spring into action at the first sign of any trouble. As was the case with other scientists targeted at EPA labs throughout the country, Dr. Long refused to cooperate in any sting operation. Because he wouldn't cooperate, the Justice Department charged Dr. Long with "benefiting from the wrongdoing of an unnamed third party." To avoid large fines that could financially cripple them for life, Dr. Long and his wife took out a second mortgage on their home and paid the Justice Department $24,000 to settle the case. (See: Nesmith J, "Focus on EPA Investigation—Cleared chemist's victory leaves bitter taste." *Atlanta Constitution*, Jan. 28, 1997, p. A6.) Similar bullying by the OIG and Justice Department was repeated at EPA research laboratories across the country as armed federal agents threatened scientists with imprisonment. After being shoved in his chair and promised five years in federal prison, one senior researcher in Duluth, Minnesota, told the Author that he planned to divorce his wife so that she could keep their home while he was in prison. Another lab director targeted in Gulf Breeze, Florida, died of a heart attack. Fortunately, federal judges eventually dismissed all of the cases working their way through the courts and the nightmare ended.

13. *Ibid, Crimes v. Kennedy Krieger*, 2001.
14. Young A, Exodus, Morale Shake CDC. *Atlanta Journal-Constitution*, 10 Sept. 2006.
15. *Atlanta Journal-Constitution*, "CDC offers buyouts in chief's office," Jul. 24, 2008.
16. Katic G, The Terry Project on CiTR #27: Silencing the Scientists. Nov. 7, 2013. http://www.terry.ubc.ca/2013/11/07/the-terry-project-on-citr-27-silencing-the-scientists/
17. Professional Institute of the Public Service of Canada (PIPSC). "The Big Chill. Silencing Public Interest Science, A Survey." http://www.pipsc.ca/bigchill.
18. *Ibid*, Oversight Hearings, 1991; Nesmith,1997.
19. Broder JM. "Title, but Unclear Power, for a New Climate Czar." *NY Times*, Dec. 11, 2008.
20. When EPA transferred the author to the Department of Marine Sciences in 1998, he assembled a group of research scientists to address the prospects of a catastrophic blowout from an offshore oil rig in the Gulf of Mexico. The approach EPA used to clean up the *Exxon Valdez* oil spill in Prince William Sound in 1989 was largely

based research that the author and his coworkers published showing that certain growth-limiting trace elements shorten the amount of time bacteria require to break down the kinds of chemicals found in crude oil (Lewis, *et al.* 1986. *Appl. Environ. Microbiol.* 51:598-603). The author's research on oil spills at UGA was aimed at developing complex mixtures of microbes capable of detoxifying crude oil and the surfactants used to disperse it in ocean water. Mixtures could include, for example, bacteria adapted to detoxifying oil at cold temperatures and high pressures near the ocean floor, as well as photosynthetic bacteria that could metabolize toxic chemicals when the oil reaches the surface. By stockpiling large quantities of freeze-dried bacteria and minerals, mixtures of oil-detoxifying microbes and trace elements could be injected directly into ruptured oil lines beneath offshore oil rigs. Wherever ocean currents carry the contaminants from there, their potential for impacting the environment would steadily diminish as the microbes detoxify them along the way. Longest, however, cut off my in-house funding and ordered local EPA managers to stop the author from collaborating with other EPA scientists (*Lewis v. EPA*, U.S. Department of Labor Case No. 98-CAA-13). EPA's Acting Asst. Administrator, Henry Longest, also denied the author access to a sophisticated computer model developed by EPA and UGA, which he needed to conduct the deepwater oil spill project.

21. Fujioka R, Vithanage G, and Yoneyama B, Analysis of proposed biosolids pellets applied to Hawaiian soil for detection and growth of Salmonella. Water Resources Research Center, University of Hawaii at Manoa, May 2004. http://www.sludgefacts.org/Ref49.pdf

22. Hyatt HM (1973). *Hoodoo—Conjuration—Witchcraft—Rootwork*, Volume II, p. 1761. Western Publishing Co., Inc., Cambridge, Maryland.

23. *Isaiah* 45:7.

24. U.S. EPA Office of Inspector General Status Report—Land Application of Biosolids, 2002-S-000004, Mar. 28, 2002. www.epa.gov/oig/reports/2002/BIOSOLIDS_FINAL_REPORT.pdf

25. Armbrister T, Weird Science at the EPA, *Reader's Digest*, Jun. 1999

26. Goldstein BD, Director EOHSI to Russo RC, Director USEPA Ecosystems Research Division, Sept. 18, 2000.

27. Russo RC to Lewis DL, Mar. 6, 2008 [Email].

28. Poland GA, Jacobson RM, The Age-Old Struggle against the Antivaccinationists. *N Engl J Med* 2011; 364:97-99. DOI: 10.1056/NEJMp1010594

Chapter 12

1. O'Dette R, Vice President, Synagro Technologies, Inc. to Stavinoha TD, Commissioner Precinct 1, Fort Bend County, TX. Nov.18, 2002 [Letter].
2. U.S. Environmental Protection Agency. 1986b. Guidelines for the Health Risk Assessment of Chemical Mixtures. EPA/630/R-98/002. Risk Assessment Forum, U.S. Environmental Protection Agency, Washington, DC. *Fed. Regist.* 51(185):34014–34025 (Sept. 24, 1986).
3. Kamler J, Soria JA. Supercritical Water Gasification of Municipal Sludge: A Novel Approach to Waste Treatment and Energy Recovery. *In*: Gasification for Practical Applications, Yongseung Y, Ed. ISBN 978-953-51-0818-4. Published: Oct. 24, 2012. DOI: 10.5772/51048.
4. Winn RN, Norris MB, Brayer KJ, Torres C, Muller SL. Detection of mutations in transgenic fish carrying a bacteriophage lambda cII transgene target. *Proc Natl Acad Sci USA*. 2000;97(23):12655-60; Winn RN, Norris MB, Lothenbach D, Flynn K, Hammermeister D, Whiteman F, et al. Sub-chronic exposure to 1,1-dichloropropene induces frameshift mutations in lambda transgenic medaka. *Mutat Res*. 2006;595(1-2):52-9; Winn RN, Majeske AJ, Jagoe CH, Glenn TC, Smith MH, Norris MB. Transgenic lambda medaka as a new model for germ cell mutagenesis. *Environ Mol Mutagen*. 2008.
5. Bove F, CDC-Atlanta, to Jennings-McElhaney J, Feb. 25, 2013 [Email].
6. Union of Concerned Scientists. Analysis on Airborne Bacteria Suppressed. Interview with James Zahn, January 2004, for UCS Scientific Integrity report. http://www.ucsusa.org/scientific_integrity/abuses_of_science/airborne-bacteria.html.
7. Marti E, Jofre J, Balcazar JL (2013) Prevalence of Antibiotic Resistance Genes and Bacterial Community Composition in a River Influenced by a Wastewater Treatment Plant. PLoS ONE 8(10):

e78906. doi:10.1371/journal.pone.0078906; Munir M, Wong K, Xagoraraki I. Release of antibiotic resistant bacteria and genes in the effluent and biosolids of five wastewater utilities in Michigan. *Water Res.* 2011 Jan;45(2):681-93. doi: 10.1016/j.watres.2010.08.033. Epub 2010 Aug 27.

8. Gurian-Sherman D, CAFOs Uncovered: The Untold Costs of Confined Animal Feeding Operations. Union of Concerned Scientists, Cambridge, MA.

9. U.S. Environmental Protection Agency. Biosolids: Targeted National Sewage Sludge Survey Report —Overview, January 2009, EPA 822-R-08-014. www.epa.gov/waterscience/biosolids/tnsss-tech.pdf

10. Cox-Foster DL, Conlan S, Holmes EC, *et al.* 2007. A Metagenomic Survey of Microbes in Honey Bee Colony Collapse Disorder. *Science* 318: 283-287.

11. Mullin CA, *et al.*, High Levels of Miticides and Agrochemicals in North American Apiaries: Implications for Honey Bee Health, 2010. *PLoS ONE* 5(March):e9754.

12. Lewis DL, Said WA. Special Applications of insect gut microflora in kinetic studies of microbial substrate removal rates. *Environ. Tox. Chem.* 1989;8:563-567; Lewis DL, 1980. Environmental and health aspects of termite control chemicals. *Sociobiol.* 5:698-703; Lewis DL, Michaels GE, Hays DB, Campbell N, Smith V. Evaluation of the anti-termitic activity of hydroxyquinoline and naphthol derivative formulations using *Reticulitermes* in laboratory and field tests. *J. Econ. Entomol.* 1978;71:818-821.

13. *Air Force Magazine*, They Flew to 65,000 Feet...The Easy Way. Vol. 47, No. 6, Jun. 1964; Moody Air Force Base, Valdosta, GA. Mar. 5, 1964 [USAF Press Release].

14. Lewis JD. Discussion provided specifically for this book, not available elsewhere.

15. Wickramasinghe C, Bacterial morphologies supporting cometary panspermia: a reappraisal. *Internat. J. Astrobiol.* 2011; 10 (1):25–30.

16. Pflug HD, Utrafine structure of the organic matter in meteorites. *In* Fundamental Studies and the Future of Science, ed. Wickramasinghe NC (1984), pp. 24–37. University College Cardiff Press, Cardiff.

17. Noffke N, Christian D, Wacey D, Hazen RM, Microbially Induced Sedimentary Structures Recording an Ancient Ecosystem in the

ca. 3.48 Billion-Year-Old Dresser Formation, Pilbara, Western Australia. *Astrobiology*. Dec. 2013, 13(12): 1103-1124. doi:10.1089/ast.2013.1030.

18. Lewis DL. Author's theories of cosmic evolution were first proposed in 1984 as an internal EPA-ORD research project in collaboration with NASA's Ames Research Center and UGA's Department of Marine Sciences.

19. Lund C, Lynch-Stieglitz J, Curry WB. *Nature*, 444. 601-604 (2006).

20. NOAA. A Paleo Perspective on Global Warming. http://www.ncdc.noaa.gov/paleo/globalwarming/paleolast.html

21. Behn MD, Lin J, Zuber MT (2004), The Thermal Structure of the Ocean Crust and the Dynamics of Hydrothermal Circulation, Geophysical Monograph 148. American Geophysical Union; Behn MD, Lin J, Zuber MT (2013), Effects of Hydrothermal Cooling and Magma Injection on Mid-Ocean Ridge Temperature Structure, Deformation, and Axial Morphology. German CR, Lin J, Parson LM, American Geophysical Union, Published Online Mar. 19, 2013. DOI: 10.1029/148GM06.

22. Transcript of Remarks by Secretary of State John Kerry. Release of Intergovernmental Panel on Climate Change Working Group 2 Report, Washington, DC, Mar. 30, 2014. http://www.state.gov/secretary/remarks/2014/03/224161.htm.

23. Transcript of Remarks by Secretary of State John Kerry. Climate Change Remarks, Jakarta, Indonesia, Feb. 16, 2014. http://www.state.gov/secretary/remarks/2014/02/221704.htm

24. See sections: "President Dwight D. Eisenhower's Farewell Address" and "President John F. Kennedy's Legacy."

25. Ibid, Kerry, Jakarta, Indonesia, 2014.

Epilogue

1. Stuck in the mud—The Environmental Protection Agency must gather data on the toxicity of spreading sewage sludge [Editorial]; Tollefson J. Raking through sludge exposes a stink. *Nature*, 2008, Vol. 453, p. 258, 262-3, May 15, 2008.

2. Perciasepe R, EPA Assistant Administrator for Office of Water, to Herman SA, EPA Assistant Administrator for Office of Enforcement

and Compliance Assurance. Request for Additional OECA Resources for the Biosolids Program. Apr. 29, 1998 [Memorandum].

3. National Research Council. Biosolids Applied to Land: Advancing Standards and Practice, p. 4. National Academy Press. Washington, DC, 2002, p. 52; [E-mail] BIRT Member R Bastian to National Academy of Sciences attaching draft of J Gaskin's final report to EPA, Mar. 13, 2001.

4. Perciasepe R, EPA Assistant Administrator, to Rominger R, Deputy Secretary, U.S. Department of Agriculture, July 24, 1997 [Letter]; U.S. House of Representatives, Committee on Science. 2000. EPA's Sludge Rule: Closed Minds or Open Debate? No. 106-95. Washington, DC: U.S. Government Printing Office.

5. NRC Panel Member Kester A to EPA officials Rubin A, Hais A, Roufael A, Carkuff A, Sajjad A, Bastian R, Brobst R, Sans C, Hamilton D, Hetherington D, Gross C, Lindsey A, Home J, Ryan J, Smith J, Colletti J, Dombrowski J, Dunn J, Walker J, Fondahl L, Dominy M, Meckes M, Murphy T. Subject: FW: Dr. David Lewis, 09/24/01 [Email]; NRC Panel Member Harrison E to Lewis DL, Mar. 5, 2003 [Email].

6. Kamler J, Soria JA. Supercritical Water Gasification of Municipal Sludge: A Novel Approach to Waste Treatment and Energy Recovery. *In*: Gasification for Practical Applications, Yun Y, Ed. ISBN 978-953-51-0818-4. Published: Oct. 24, 2012. DOI: 10.5772/51048.

7. U.S. Environmental Protection Agency, Technical Assistance Directory, Office of Research and Development. EPA/600/K-97/001 October 1997.

8. University of Georgia, Bush taps former UGA dean for REE under secretary. Jan. 19, 2006.

9. *Lewis v. EPA, U.S. Department of Labor, CA 2003-CAA-00005, 00006.* Deposition of RE Hodson, Ph.D. Jan. 31, 2003. Also, see *USA, ex rel. Lewis et al. v. Walker et al.* U.S. District Court, Middle District of Georgia, Athens Division. Case No. 3:06-CV-16-CDL, Affidavit of Guthrie L, and the following testimony of Lewis DL: *Dr. Larry Guthrie, head of the Dairy Science Department, investigated problems at Relator Boyce's dairy farm and supported the conclusions put forth by Relator Boyce and his experts, namely, that long-term application of sewage sludge to Relator Boyce's dairy farm had contaminated the land*

with hazardous wastes such that forage crops were too toxic to feed to dairy cattle even if they were diluted with uncontaminated forage and cattle feed were amended with copper. On the one hand, Dr. Gale Buchanan, Dean of the College of Agricultural and Environmental Sciences, prohibited Dr. Guthrie from testifying as an expert for Relators Boyce and McElmurray in their lawsuits against Augusta. On the other hand, Dean Buchanan allowed Defendants Gaskin, Miller and other UGA Defendants conduct a study with Defendant Brobst, who worked directly with attorneys defending Augusta. [Author's Note: EPA's Cooperative Agreement with the Water Environment Federation (Coop. Agr. CX820725-01-2 titled "National Biosolids Public Acceptance Campaign"), which was funded with Congressional earmarks, specifically mentions using deans at land grant universities as gatekeepers to keep track of research projects on biosolids and identify scientists who oppose land application of biosolids.]

10. Lewis DL, Institutional Research Misconduct: An Honest Researcher's Worst Nightmare, *Autism Sci Digest*, 2012(4):31-40.
11. Beecher N, David Lewis, PhD, Shifts Focus. *NEBRA*, Apr. 9, 2012. www.nebiosolids.org/.
12. Beecher N, Synagro responds to biosolids recycling opponent. *New England Water and Wastewater News*, Issue 69, Nov. 2001, p. 7.
13. *NEBRA*, Mission and Membership. http://www.nebiosolids.org/.
14. *Ibid.*

Afterword

1. Egilman DS, Bohme S (2005), Over a Barrel: Corporate Corruption of Science and the Effects on Workers and the Environment. *Int J Occup Environ Health* 11: 331-337; Krimsky S. (2003), Science in the Private Interest: Has the Lure of Profits Corrupted Biomedical Research? Rowman & Littlefields Publishers Inc. Lanham- Boulder- New York- Oxford; Michaels D (2008), Doubt is Their Product: How Industry's Assault on Science Threatens Your Health. Oxford University Press. New York.

2. Michaels (op cit.) in his book exposing the tobacco industry's campaign to cover up health hazards, states "every chapter in my book contains material that was uncovered during the discovery process in a legal proceeding."

3. *The Merchant of Venice* 2.7.7-10

4. Caroline Snyder holds a Ph.D. from Harvard University and is Professor Emeritus at the Rochester Institute of Technology, where she developed and taught interdisciplinary environmental science courses. Before retiring, Dr. Snyder chaired the Department of Science, Technology, and Society. For the past seventeen years she has researched and written about the politics and science of using municipal sewage sludge (aka biosolids). Her article published in 2002 in the *International Journal of Occupational and Environmental Health*, titled "The Dirty Work of Promoting 'Recycling' of America's Sewage Sludge," is widely cited.

Appendix I.

1. Bastian RK, Interpreting Science in the Real World for Sustainable Land Application 2005; *JEQ* 34,1:174.
2. EPA Fact Sheet. http://water.epa.gov/polwaste/wastewater/treatment/biosolids/
3. Hale RC, LaGuardia MJ, Harvey EP, Gaylor MO, Mainor TM, Duff WH. Persistent pollutants in land applied sludges. *Nature* 412:140-141.
4. NEBRA, Response to Toxic Action Center's Toxic Sludge in Our Communities. Mar. 3, 2003.
5. CalRecycle. http://www.calrecycle.ca.gov/organics/biosolids/
6. Gattie DK, Lewis DL, A high-level disinfection standard for land-applied sewage sludge (biosolids). *Environ. Health Perspect.* 2004;112:126-31.
7. Gibbs RA *et al.* Re-growth of faecal coliforms and salmonellae in stored biosolids and soil amended with biosolids. *Water Science and Technology* 1997; 35 (11-12).
8. Miles SL, Takizawa K, Gerba CP, Pepper IL, Survival of Infectious Prions in Blass B Biosolids. *J.Env..Sci. & Hlth.* 2011; 46: 364-370.
9. Kaplan N, Prions' Great Escape. http://www.nature.com/news/2008/080701/full/news.2008.926.html
10. Toffey WE, Biosolids Odorant Emissions as a Cause of Somatic Disease. Presentation to the 2007 North East Bisolids & Residuals Conference, Philadelphia Water Department. December 4, 2007.
11. Shusterman D, Critical review; the health significance of environmental odor pollution. *Arch.Environ.Health* 1992; 47:76-87.
12. NEBRA Mar. 3, 2003 op.cit p. 10.

13. Lewis DL *et al.*, Interactions of pathogens and irritant chemicals in land-applied sewage sludges (biosolids) BMC 2002; 2:11. http://www.biomedcentral.com/1471-2458/2/11; Lewis DL, Gattie DK, Pathogen risks from applying sewage sludge to land *Environ. Sci. Technol.* 2002; 36:286A-293A; Ghosh J, Bioaerosols Generated From Biosolids Applied Farm Fields In Wood County, Ohio. Master of Science Thesis (2005), Graduate College of Bowling Green State University. Abstract by Robert K Vincent, Advisor. www.ohiolink.edu/etd/sendpdf.cgi/Ghosh%20Jaydeep.pdf?bgsu1131322484; Khuder S *et al. Arch. Environ. Occup. Health* 2007; 62, 5–11.

14. NEBRA. Mar. 3, op.cit. p. 22.

15. U. S. Environmental Protection Agency. Biosolids: Targeted National Sewage Sludge Survey Report - Overview, January 2009, EPA 822-R-08-014.   http://water.epa.gov/scitech/wastetech/biosolids/tnsss-overview.cfm; See also Sepulvado JG, Blaine AC, Hundal LS, Higgins CP, Occurrence and Fate of Perfluorochemicals in Soil Following the Land Application of Municipal Biosolids. *Environ. Sci. Technol.* Mar. 29, 2011, DOI: 10.1021/es103903d

16. Lewis DL, Garrison W, Wommack KE, Whittemore A, Steudler P, Melillo J. Influence of environmental changes on degradation of chiral pollutants in soils. *Nature* 1999; 401:898-901; Paris DF, Lewis DL. Chemical and microbial degradation of ten selected pesticides in aquatic systems. *Residue reviews* 1973; 45:95-124.

17. Abernethy MD, To sludge or not to sludge?: At summit, scientists discuss risks, Interview with R Chaney, USDA. Green Consumer Headlines, *Times-News*, May 2, 2010. http://www.managemylife.com/mmh/articles/curated/278108

18. U.S. Environmental Protection Agency, Report: EPA-600/S1-81-026, 232 p. (Apr. 1981). Sewage Sludge – Viral and Pathogenic Agents in Soil-Plant-Animal Systems. G.T. Edds and J.M. Davidson, Institute of Food and Agricultural Systems, University of Florida. An EPA Project Summary is available at http://nepis.epa.gov/ by searching 600S181026 or key words in the title of the report.

19. *Ibid*, U.S. EPA Office of Inspector General Status Report (2002).

20. NEBRA, Is biosolids recycling safe? How do we know? http://www.nebiosolids.org/index.php?page=faqs

21. Harrison EZ, McBride MB, Bouldin DR, Land application of sewage sludges: an appraisal of the US regulations. *Int. J.Environment and Pollution*, Vol. 11, No.1. 1-36. Retrieved at http:cwmi.css.cornell.edu/PDFS/LandApp.pdf. See also Case for Caution Revisited 2008 (revised 2009), http:cwmi.css.cornell.edu/case.pdf; http://cwmi.css.cornell.edu/PDFS/LandApp.pdf

22. National Academy of Sciences, National Research Council. Biosolids Applied to Land: Advancing Standards and Practices, National Academy Press, Jul. 2, 2002. www.nap.edu/books/0309084865/html

Appendix III.

1. Billups Grove Baptist Church is an historic African-American church near Athens, Georgia, which was organized in 1898 by former slaves. In 1976, the City of Athens created a landfill near the church, and now composts sewage sludge (biosolids) generated on land directly behind the church. Families attending the church, and living around it, suffer from exposure to landfill leachates and sewage sludges, which contaminate the air, water and soil. Symptoms include those commonly reported in the scientific literature at other land-application sites, such as gastrointestinal problems and chronic infections of the skin and respiratory tract.

2. In 1977, President Jimmy Carter announced a 10-year program to construct wastewater (sewage) treatment plants for every municipality in the country. In a speech to Congress, he cautioned: "But at the same time, we need to be sure that sewage projects supported by Federal money do not create additional environmental problems... We also must ensure that the systems are operated properly...that there is an effective pretreatment program to remove harmful industrial wastes from these systems; and that we are carefully considering alternative solutions...". President Jimmy Carter, The Environment Message to the Congress. May 23, 1977. www.presidency.ucsb.edu/ws/index.php?pid=7561

Appendix IV.

1. WEF/EPA Biosolids Fact Sheet (1997) titled "Biosolids Reuse in Southern California" funded by EPA Cooperative Agreement #CX820725-01-2, National Biosolids Public Acceptance Campaign.
2. EPA-WEF National Biosolids Public Acceptance Campaign, Grant No. CX-820725-01-0 and renewals, Decision Memorandum from Michael J. Quigley to Michael B. Cook, July 28, 1992.
3. Suquamish Tribe of Port Madison, WA. *Chief Seattle's 1854 Oration.* www.suquamish.nsn.us/HistoryCulture/Speech/tabid/85/Default.aspx

Appendix V.

1. In 2008, the Senate Environment Public Works Committee scheduled, then cancelled, a Briefing on EPA's biosolids program. Chapter 5 of this book contains a detailed discussion of the planned Briefing and reasons it was cancelled. Appendices V and VI represent the written testimony of the Author and former dairy owner Andy McElmurray, respectively. Mr. McElmurray, another dairy owner, William (Bill) Boyce, and their families sued the City of Augusta and Richmond County over damages caused by hazardous wastes in Augusta's sewage sludge. The McElmurray family also filed suit against the USDA to obtain compensation for crop losses caused by the contamination of their land. They won all of their cases based on compelling evidence that EPA, the Georgia Environmental Protection Division, and the City of Augusta covered up the adverse effects Augusta's biosolids had on their dairy cattle. The cover-up included publishing environmental monitoring data fabricated by the City of Augusta, which indicated levels of heavy metals and nitrogen in Augusta's sewage sludges generally dropped to safe levels after EPA passed its current sludge regulation (503 sludge rule) in 1993.
2. U.S. Environmental Protection Agency. Memorandum: Walker JM, EPA Municipal Technology Branch to Henry L. Longest II, EPA Assoc. Deputy Asst. Administrator for Water Program Operations. Sept. 12, 1978.
3. Henry Longest was Acting Asst. Administrator of EPA's Office of Research & Development (ORD) during the time Dr. Lewis

researched biosolids at UGA; and Mr. Longest was deposed in Dr. Lewis' U.S. Department of Labor (DOL) Cases. The EPA lab where Dr. Lewis worked played a lead role in peer-reviewing EPA's sludge rule in 1992; and Longest's lead role in promoting biosolids and over-seeing the development of EPA's sludge regulations was common knowledge throughout EPA-ORD. EPA Attorney Bridget Shea, however, informed Dr. Lewis' attorneys prior to his DOL depositions that Mr. Longest would not answer any questions related to biosolids.

4. Lewis, DL, EPA science: Casualty of election politics. *Nature* 1996; 381:731-2.

5. Lewis, DL, Arens M, Appleton S, *et al.*, Cross-contamination potential with dental equipment. *Lancet* 1992;340:1252-4; Lewis DL, Arens M, Resistance of microorganisms to disinfection in dental and medical devices. *Nature Medicine* 1995;1:956-8; CDC, Recommended infection-control practices for dentistry. *Morbid. Mortal. Wky. Rep.* 1993; 42:1-12.

6. Mikhail N, Lewis DL, Omar N, Taha H, El-Badawy A, Mawgoud nA, Abdel-Hamid M, Strickland GT, Prospective study of cross-infection from upper-GI endoscopy in a hepatitis C–prevalent population. *Gastrointest Endosc* 2007; 65:584-588.

7. U.S. EPA Office of Inspector General Status Report—Land Application of Biosolids, 2002-S-000004, Mar. 28, 2002. www.epa. gov/oig/reports/2002/BIOSOLIDS_FINAL_REPORT.pdf

8. Harrison E. Correspondence at Nature.com. June 17, 2008.

9. Tollefson J. Raking through sludge exposes a stink. *Nature*, 2008, Vol. 453, p. 258, 262-3, May 15, 2008.

10. *Nature*, Stuck in the Mud. Vol. 453 p. 258, May 15, 2008.

11. U.S. House of Representatives, Committee on Science. 2000a. EPA's Sludge Rule: Closed Minds or Open Debate? No. 106-95. Washington, DC: U.S. Government Printing Office; 2000b. Intolerance at EPA—Harming People, Harming Science? No. 106- 103. Washington, DC: U.S. Government Printing Office.

12. Notification and Federal Employee Antidiscrimination and Retaliation Act of 2002. 2002. Public Law 107-174.

13. Deposition transcript of EPA-Athens Research Director Dr. Robert Swank, p. 52. Sep. 6, 2000. *Lewis v. EPA*, U.S. Department of Labor Case Nos. 99-CAA-12, 2000-CAA-10, 2000-CAA-11.

14. Kuehn R., Suppression of Environmental Science, *American Journal of Law & Medicine*, 2004;30:333-69.

15. *McElmurray v. USDA*, United States District Court, Southern District of Georgia, Case No. CV105-159, Order issued Feb. 25, 2008.

16. Lewis DL, Garrison W, Wommack KE, Whittemore A, Steudler P, Melillo J, Influence of environmental changes on degradation of chiral pollutants in soils. *Nature* 1999; 401:898-901.

17. *Union of Concerned Scientists*, Cambridge, MA. Hundreds of EPA scientists report political interference over last five years. Apr. 23, 2008. (www.ucsusa.org).

18. Gattie DK, Lewis DL, A high-level disinfection standard for land-applied sewage sludge (biosolids). *Environ. Health Perspect.* 2004;112:126-31.

19. Griffin DW, Atmospheric Movement of Microorganisms in Clouds of Desert Dust and Implications for Human Health. *Clin. Microbiol. Rev.* Jul 2007; 20(3): 459–477.

20. *Ibid.*

21. Lewis DL, Gattie DK, Novak ME, Sanchez S, Pumphrey C. Interactions of pathogens and irritant chemicals in land-applied sewage sludges (biosolids). *BMC Public Health* 2:11 (28 Jun). www.biomedcentral.com/1471-2458/2/11.

22. Dorn RC, Reddy CS, Lamphere DN, Gaeuman JV, Lanese R: Municipal sewage sludge application on Ohio farms: health effects. *Environ. Res.* 1985; 38:332-359.

Appendix VI.

1. Based on independent analyses performed by the Georgia Environmental Protection Division and a private consulting firm, potentially harmful levels of thallium and other hazardous wastes were found in milk samples collected on one dairy farm treated with Augusta's sewage sludge, and in milk cartons pulled from the shelf in area grocery stores. *USA, ex rel. Lewis, McElmurray & Boyce et al. v. Walker et al.* Case No. 3:06-CV-16. Lewis DL, Responses to Interrogatories submitted by Defendant UGA Research Foundation: "Milk Contamination Cover-Up," p. 145-149. May 4, 2009.

2. *McElmurray v. United States Department of Agriculture*, United States District Court, Southern District of Georgia. Case No. CV105-159. Order issued Feb. 25, 2008.

3. *Ibid.*

4. *USA, ex rel. R.A. McElmurray, III, G. William Boyce, and David L. Lewis v. The Consolidated Government of Augusta-Richmond County, Georgia.* United States District Court for the Northern District of Georgia, Atlanta Division. Civil Action No. 1:05-CV-1575-ODE.

5. *Ibid, McElmurray v. USDA,* Order issued Feb. 25, 2008.

6. *Ibid.*

7. Bastian R, EPA Office of Water, to Dominy M, EPA Region IV-Atlanta, Nov. 25, 2003 [Email]. Bastian states: "I have been drafted by OST to develop a write-up on the Augusta, GA, case to include in the petition response." Bastian attached a draft "write-up" and explained that it incorporated suggestions from Bob Brobst. Brobst coauthored the 2003 *JEQ* study with J. Gaskin. This write-up was incorporated in a letter to Joseph Mendelson, III and Thomas Linsey issued by EPA Asst. Administrator G. Tracy Mehan, III on Dec. 24, 2003; Mehan, GT III, EPA Asst. Administrator for Water, to Mendelson J III, Legal Director, Center for Food Safety, Linsey T, Community Environ. Legal Defense Fund, Dec. 24, 2003. This letter dismissed in link between biosolids and the deaths of Tony Behun, Shayne Conner, Daniel Pennock, and the dairy cattle on the McElmurray and Boyce farms.

8. Deposition testimony of former Augusta Wastewater Treatment Plant Manager Allen Saxon taken on July 23, 1999 (p. 166), and deposition of Hugh Avery taken on May 6, 1999 (p. 19). McElmurray et al. and Boyce et al. v. Augusta, Georgia, Richmond County Superior Court, Civil Action File Nos. 198-216, 198-217.U.S. EPA Office of Inspector General Status Report - Land Application of Biosolids, 2002-S-000004, Mar. 28, 2002.

9. BIRT Member R Bastian to National Academy of Sciences attaching draft of J Gaskin's final report to EPA, Mar. 13, 2001.

10. Gaskin J, Brobst R, Miller W, Tollner W. 2003. Long-term biosolids application effects on metal concentrations in soil and bermudagrass forage, Table 2. *J. Environ. Qual.* 32: 146-152. http://jeq.scijournals.org/cgi/reprint/32/1/146.pdf

11. *Ibid,* Bastian, 2001.

12. Holmes C, University of Georgia. Sludge study relieves environmental fears. Jan. 29, 2003.

13. *Ibid,* Bastian, 2003.

14. *Ibid*, Mehan, 2003.
15. U.S. EPA Office of Inspector General Status Report—Land Application of Biosolids, 2002-S-000004, Mar. 28, 2002.
16. *Lewis v. EPA*, U.S. Department of Labor, CA 2003-CAA-00005, 00006. Deposition of Robert E. Hodson, Ph.D. Jan. 31, 2003. Prof. Hodson, Director of Marine Programs at UGA, testified that UGA's Provost and other administrators advised him not to hire Dr. Lewis after EPA's Office of Water funded the Gaskin study at UGA. The reason, Prof. Hodson testified, was "because we're dependent on this money...grant and contract money...money either from possible future EPA grants or [from] connections there might be between the waste-disposal community [and] members of faculty at the university."

INDEX